T0191601

Distributed Graph Analytics

Unnikrishnan Cheramangalath • Rupesh Nasre •
Y. N. Srikant

Distributed Graph Analytics

Programming, Languages, and Their Compilation

 Springer

Unnikrishnan Cheramangalath
Department of Computer Science
and Engineering
Indian Institute of Technology Palakkad
Palakkad, Kerala, India

Rupesh Nasre
Department of Computer Science
and Engineering
Indian Institute of Technology Madras
Chennai, India

Y. N. Srikant
Department of Computer Science
and Automation
Indian Institute of Science
Bangalore, India

ISBN 978-3-030-41888-5   ISBN 978-3-030-41886-1   (eBook)
https://doi.org/10.1007/978-3-030-41886-1

This Springer imprint is published by the registered company Springer Nature Switzerland AG.
The registered company address is: Gewerbestrasse 11, 6330 Cham, Switzerland

Preface

Graphs model several real-world phenomena. Graph algorithms have been an interesting area of study and research due to rich structural information embedded into graphs. However, it is only in the last couple of decades that systemic aspects of graph algorithms have been pursued more prominently. Especially with the advent of large networks, and storage of huge amount of unstructured data, efficient graph processing has become a crucial piece in the performance puzzle of the underlying applications. For example, efficient graph processing is central to detection of online frauds, improvements in advertising and business, study of how diseases spread, computational geometry, and natural language processing, to name a few.

In parallel, we witnessed prominent innovations in computer architecture. Shared memory systems with a large number of cores, NUMA architecture, accelerators such as GPUs and FPGAs, and large decentralized systems are so common that high-performance computing is no longer a niche area. Almost every field, be it physics, biology, or material science, need to depend on such architectures for efficient simulation of a natural phenomenon. On the industry front, data warehouses are becoming bulky with pressures to store, maintain, and analyze tremendous amount of structured and unstructured data. Graph databases, textgraphs, and their modeling as hypergraphs pose new challenges. Processing such multi-modal data requires different algorithms. In addition, an algorithm needs to be tuned for a certain backend hardware for efficient implementation, because an optimized implementation for one hardware may not be efficient on the other.

Our goal in this book is to marry these two important trends: graph algorithms and high performance computing. Efficient and scalable execution of graph processing applications demands innovations at multiple levels: algorithm, its associated data structures, their implementation, and their implementation tuned to a certain hardware. Managing complexity at so many levels of abstraction is clearly a challenge for users. The issue gets exacerbated as domain experts may not be HPC experts, and vice versa. Therefore, the role of programming languages and the associated compilers is crucial to automate efficient code generation for various architectures. This book pens down essentials on these aspects.

We divide this book in three parts: programming, languages, and their compilation. The first part deals with manual parallelization of graph algorithms. It uncovers various parallelization patterns we encounter while dealing especially with graphs. The second part exploits these patterns to provide language constructs using which a graph algorithm can be specified. A programmer can work only with those language constructs, without worrying about their implementation. The implementation, which is the heart of the third part, is handled by a compiler, which can specialize code generation for a backend device. We present a few suggestive results on different platforms, which justify the theory and practice covered in the book. Together, the three parts provide the essential ingredients in creating a high-performing graph application.

While this book is pitched at a graduate or advanced undergraduate level as a specialized elective in universities, to make the matter accessible, we have also included a brief background on elementary graph algorithms, parallel computing, and GPUs. It is possible to read most of the chapters independently, if the readers are familiar with the basics of parallel programming and graph algorithms. To highlight recent advances, we also discuss dynamic graph algorithms, which pose new challenges at the algorithmic and language levels, as well as for code generation. To make the discussion more concrete, we use a case study using Falcon, a domain-specific language for graph algorithms. The book ends with a section on future directions which contains several pointers to topics that seem promising for future research.

We believe this book provides you wings to explore the exciting area of distributed graph analytics with zeal, and that it would help practitioners scale their graph algorithms to newer heights.

We would like to especially thank Ralf Gerstner, Executive Editor of Computer Science at Springer-Verlag for his encouragement and patient advice.

Palakkad, India   Unnikrishnan Cheramangalath
Chennai, India   Rupesh Nasre
Bangalore, India   Y. N. Srikant
January 2020

Contents

Chapter 1
Introduction to Graph Analytics

This chapter discusses the importance of graph analytics. It describes important concepts in elementary graph theory and different graph representations, programming frameworks for parallelization, and various challenges posed by graph analytics algorithms. Graph partitioning and real-world applications of graphs are also covered. Frameworks and DSLs that can ease programming graph analytics are briefly discussed.

1.1 What Is Graph Analytics?

Graph analytics or graph analysis involves usage of graph algorithms to determine relationships between objects in a graph and also its overall structural characteristics. While traditional applications of graph analytics such as compiler optimizations, job scheduling in operating systems, database applications, natural language processing, computational geometry etc. continue to be important, a major business driver of this area in recent times has been social network analysis which identifies potential "network influencers" who, in turn, influence buying products and services by social network communities. However, many more important applications of graph analytics have been proposed of late:

- Detecting fraudulent transactions in banking, fraud in insurance claims, financial crimes such as money laundering, illegal activities in telecommunication networks, etc.
- Molecular simulations, disease identification, etc.
- Optimizing supply and distribution chains, routes in airline networks, transportation networks, communication networks, etc.
- Power and water grid analysis.
- Prevention of cyber attacks and crimes in computer networks.
- Social network analysis for use in business, marketing, etc.

© Springer Nature Switzerland AG 2020
U. Cheramangalath et al., *Distributed Graph Analytics*,
https://doi.org/10.1007/978-3-030-41886-1_1

Examples of different kinds of graph analysis that are used in various applications are path analysis, connectivity analysis, clustering, etc. Path analysis is used to find the shortest distance between pairs of nodes in a graph and is useful in optimizing transportation networks supply and distribution chains, etc. Connectivity analysis helps in determining whether parts of networks are disconnected or connected by too few links, and is useful in power and water grid analysis, telecommunication network analysis, etc. Clustering helps in identifying groups of interacting people in a social network. Other types of analyses that use graph analytics are community analysis and centrality analysis.

While graphs in traditional applications are small or can fit in the memory of a single desktop of server box, social applications of graph analytics usually involve massive graphs that cannot fit into the memory of a single computer. Such graphs can be handled using a distributed computation. The graphs are partitioned and distributed across computing nodes of a distributed computing system, and computation happens in parallel on all the nodes. Efficient implementation of graph analytics on such platforms requires a deep knowledge of the hardware, threads, processes, inter-process communication, and memory management. Tuning a given graph algorithm for such platforms is therefore quite cumbersome, since the programmer needs to worry not only about *what* the algorithmic processing is, but also about *how* the processing would be implemented. Programming on these platforms using traditional languages along with message passing interface (MPI) and tools is therefore challenging and error-prone. There is little debugging support either.

To exploit the computational power of a single machine, graph analytic algorithms are parallelized using multiple threads. Such a multicore processing approach can improve the efficiency of the underlying computation. Unlike in a distributed system, all the threads in a multicore system share a common memory. A multicore system is programmed using OpenMP and libraries such as pthreads.

Apart from distributed systems and multicore processors, Graphical Processing Units (GPUs) are also widely used for general purpose high performance computing. GPUs have their own cores and memory, and are connected to a CPU via a PCI-express bus. Due to massive parallelism available on GPUs, threads are organized into a hierarchy of warps and thread blocks. All the threads within a warp execute in single-instruction multiple data (SIMD) fashion. A group of warps constitutes a thread block, which is assigned to a multi-processor for execution. GPUs are programmed using programming languages such as CUDA and OpenCL. Graph algorithms contain enough parallelism to keep thousands of GPU cores busy.

As noted above, one needs to use different programming paradigms for parallelizing a given application on different kinds of hardware. As data sizes grow, it is imperative to combine the benefits of these hardware types to achieve best results. Thus, both the graph data as well as the associated computation would be split across multi-core CPUs and GPUs, operating in a distributed manner. It is an enormous effort for the programmer to optimize the computation for three different platforms in multiple languages. Debugging such a parallel code at large scale is also often practically infeasible.

This problem can be addressed by graph analytic frameworks and domain specific languages (DSL) for graph analytics. A DSL hides the hardware details from a programmer and allows her/him to concentrate only on the algorithmic logic. A DSL compiler should generate efficient code from the input DSL program. This book takes a deep look at various programming frameworks, domain specific languages, and their compilers for graph analytics for distributed systems with CPUs and GPUs.

1.2 Graph Preliminaries

A graph G(V, E) consists of a set of vertices $V = \{v_1, v_2, \ldots, v_n\}$ and a set of edges $E = \{e_1, e_2, \ldots, e_m\}$. An edge $e \in E$ consist of two vertices $u, v \in V$. The edge set $E \subseteq \{(u, v)|u, v \in V\}$. The *order* and *size* of a graph are the cardinality of its vertex set and edge set respectively. The vertices and edges in a graph may be attached with optional mutable properties and these properties relate to the real-life application that the graph is used in. A graph is called *weighted* if every edge is associated with an integer value (positive, negative, or zero).

A graph is *undirected* if its edges are *unordered* pairs, and in such a case, edges are considered bidirectional. A graph is *directed* if its edges are ordered pairs. Two vertices are *adjacent* if there is an edge between them. For a *directed* edge e: $u \rightarrow v$, e is directed from u to v. The vertices u and v are called *tail* and *head* respectively. The vertices u and v are also addressed with the names *source* and *sink* respectively. An edge is called a *self loop* if its *tail* and *head* vertices are the same, otherwise, often it is called a *link*.

A *path* in a directed graph is a finite sequence of edges which connect vertices and no vertex is repeated, except possibly for the first and the last vertices which may be the same. A path in a directed graph is also *directed*. That is, $E_p = \{e_1, e_2, \ldots, e_k\} \subseteq E$ is a *directed path* if

$$E_p \subseteq E \wedge \{v_1, v_2, \ldots, v_{k+1}\} \subseteq V \wedge e_i : v_i \rightarrow v_{(i+1)}, 1 \leq i < k \qquad (1.1)$$

A similar definition holds for undirected graphs as well, but obviously the edges are not directed. A graph is *simple* if it does not have a loop and there are no multiple edges between two vertices. A graph is *complete* if every pair of distinct vertices is connected by a unique edge.

For undirected graphs, the *degree* of a vertex is the number of edges incident on the vertex. For a directed graph, the *indegree* of a vertex v is the number of edges having vertex v as their heads. Similarly, the *outdegree* of a vertex v is the number of edges having vertex v as their tail.

A graph G_s (V_s, E_s) is a subgraph of the graph G (V, E) if $V_s \subseteq V$ and $E_s \subseteq E$. G_s (V_s, E_s) is a proper subgraph of the graph G (V,E) if $V_s \subset V$ or $E_s \subset E$. A directed graph is *strongly connected* if there is at least one *directed path* between any pair of distinct vertices. A *strongly connected component* (SCC) of a directed graph G is a maximal strongly connected subgraph of G. All SCCs are disjoint and none of the SCCs can be extended with more vertices and edges from G, still retaining strong connectivity. An undirected graph is *connected* if there is at least one *path* between any pair of distinct vertices. A *connected component* (CC) of an undirected graph G is a maximal connected subgraph of G. All CCs are disjoint, and none of the CCs can be extended with more vertices and edges from G, still retaining connectivity.

Distance between two vertices is the length of the shortest path between them. The *diameter* of a graph is defined as the largest distance between any pair of vertices. A *cycle* of a graph G(V,E) is a set $E_0 \subseteq E$ that forms a *path* such that the first and the last vertices of the *path* are the same. A graph is *cyclic* if it has at least one *cycle*; otherwise it is *acyclic*. A *tree* is an *undirected*, *connected* and acyclic graph. For a tree, $|V| = |E| + 1$. A *rooted* tree is a tree which has a distinguished vertex called the *root*, and all edges are oriented away from the root. Usually, a rooted tree is referred to as a tree and the root is identified, with the directions of the edges not being explicitly mentioned. A *spanning* tree for an undirected connected graph G is a tree that connects all the vertices of G. If G is disconnected, then we talk about a *spanning forest* for G, with one spanning tree for each connected component of G.

An undirected graph is *bipartite* if the vertices in the graph can be divided into two unique and independent subsets such that for all the edges the two endpoints do not belong to the same subset. That is, a graph G(V, E) is bipartite if

$$\exists P, Q \subset V \wedge P \cup Q = V \wedge P \cap Q = \phi \wedge e : (u, v) \in E \implies$$

$$u \in P, v \in Q \parallel u \in Q, v \in P \qquad (1.2)$$

Figure 1.1 shows examples of undirected graphs: a complete graph (Fig. 1.1a), a connected graph (Fig. 1.1b), a bipartite graph (Fig. 1.1c), and a tree (Fig. 1.1d).

Figure 1.2 shows directed graphs with and without edge weights, and a strongly connected graph.

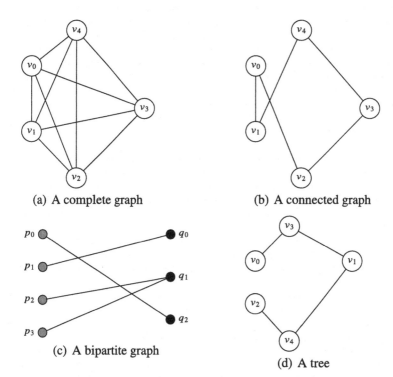

(a) A complete graph

(b) A connected graph

(c) A bipartite graph

(d) A tree

Fig. 1.1 Samples of undirected graphs

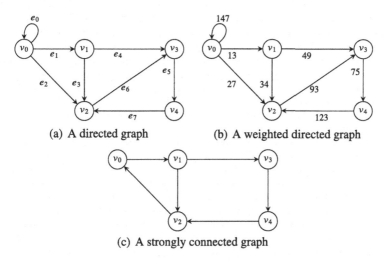

(a) A directed graph

(b) A weighted directed graph

(c) A strongly connected graph

Fig. 1.2 Samples of directed graphs

1.3   Graph Storage Formats

Graphs are stored in memory in various formats. We describe these formats below.

1.3.1   Adjacency Matrix

The adjacency matrix *Arr* of a graph G(V, E) is a square matrix of size $|V| \times |V|$ such that:

$Arr[i][j] = \text{weight}(e(i,j))$, if $e(i, j) \in E$
$Arr[i][j] = \infty$ if $i \neq j \land e(i, j) \notin E$.
$Arr[i][j] = 0$ if $i = j \land e(i, j) \notin E$

Table 1.1 shows the adjacency matrix representation for the graph in Fig. 1.2b. For undirected graphs, the corresponding adjacency matrix is symmetric, that is, Arr[i][j] = Arr[j][i].

1.3.2   Compressed Sparse Row

The Compressed Sparse Row (CSR) format represents an unweighted graph using a table of two rows and a weighted graph with a table of three rows. Table 1.2 shows the CSR representation for the directed graph in Fig. 1.2b. The third row named *weight* stores the edge weight for each edge, and has as many elements as the number of edges in the graph. The first row named *index* stores pointers into the row *weight* and has $|V| + 1$ number of entries. The second row stores the destination vertex (*head*) of the edges in sorted order. Elements in the row *weight*, between the offsets *index*[i] and *index*[i + 1] store weights of all the edges with v_i as the *tail* vertex. For example, in Table 1.2, edge weights for edges having vertex v_1 as tail

Table 1.1 Adjacency matrix representation of the graph in Fig. 1.2b

	v_0	v_1	v_2	v_3	v_4
v_0	147	13	27	∞	∞
v_1	∞	0	34	49	∞
v_2	∞	∞	0	93	∞
v_3	∞	∞	∞	0	75
v_4	∞	∞	123	∞	0

Table 1.2 Compressed sparse row representation of the graph in Fig. 1.2b

	v_0	v_1	v_2	v_3	v_4			
Index	0	3	5	6	7	8		
Head vertex	v_0	v_1	v_2	v_2	v_3	v_3	v_4	v_2
Weight	147	13	27	34	49	93	75	123

start at index 3 in row *weight*, and there are two such edges ($index[v_2] - index[v_1] = 5 - 3 = 2$). It is to be noted that if a vertex v_i is isolated (no edges connected to it), then $index[v_i] = index[v_{i+1}]$. The last entry in the row *index* is always equal to $|E|$ and is useful in computing the number of edges emanating from the vertex $index[v_{last}]$. Unweighted graphs do not need the *weight* row. Undirected graphs can be stored in CSR format by storing each edge in both the directions. The space complexity of the CSR format is O(V + E). This storage representation is very efficient for sparse graphs, i.e., $|E| << |V|^2$.

1.3.3 Incidence Matrix

The incidence matrix representation of a graph with p vertices and q edges is a two dimensional array A, with p rows and q columns. For an *undirected* graph, $A[v_x, e_y] = 1$, if the edge e_y is incident on vertex v_x; otherwise 0. For a *directed* graph, $A[v_x, e_y] = 1$ if *tail* $(e_y) = v_x$; -1 if *head* $(e_y) = v_x$; and 0 if e_y is not connected to vertex v_x. The incidence matrix for the graph in Fig. 1.2a is shown in Table 1.3.

1.3.4 Other Formats

Edge-List In an *edge-list* format, each edge e($src \xrightarrow{weight} dst$) is stored as a triple (*src, dst, weight*) sorted by the source vertex-id. Storing each edge in its reverse form as a triple (*dst, src, weight*) sorted by the destination vertex-id is useful for *pull*-based algorithms wherein a vertex receives updates on its incoming edges. The storage requirement for both these formats is $3 \times |E|$, which is greater than that of the CSR format.

Adjacency List This is a data structure with an array of separate lists, one for each vertex. Each array element A_i points to a list of vertices v_j adjacent to v_i. If the graph is weighted, then the weight of the edge (v_i, v_j) is also stored along with the element v_j, in the list of the array element A_i (corresponding to vertex v_i). The space complexity of adjacency list representation is O($|V| + |E|$). A comparison

Table 1.3 Incidence matrix representation for the graph in Fig. 1.2a

	e_0	e_1	e_2	e_3	e_4	e_5	e_6	e_7
v_0	1	1	1	0	0	0	0	0
v_1	0	−1	0	1	1	0	0	0
v_2	0	0	−1	−1	0	0	1	−1
v_3	0	0	0	0	−1	1	−1	0
v_4	0	0	0	0	0	−1	0	1

Table 1.4 Comparison of graph storage formats

Storage	Space complexity	Access time
Adjacency matrix	$O(|V|^2)$	$O(1)$
CSR	$O(|V|+|E|)$	$O(\text{vertex-degree})$
Edge-list	$O(|E|)$	$O(\text{vertex-degree})$
Adjacency list	$O(|V|+|E|)$	$O(|V|+\text{vertex-degree})$
Incidence matrix	$O(|V|^2)$	$O(1)$

of the graph storage formats is given in Table 1.4, where *Access Time* is the time required to check the presence or absence of an edge.

1.4 Graph Partitioning Strategies

When a graph is too large to fit in the memory of a single computer, distributed computation is required. The graph is partitioned into subgraphs and these are distributed across the computing nodes of a cluster (distributed computing system). More formally, a graph G(V, E) is partitioned into k subgraphs $G_1, G_2, \ldots, G_k$, where $k \geq 2$. The vertices and edges of the subgraphs G_i can be chosen in several ways. The best partition must have almost equal number of vertices in each subgraph and the number of inter-partition edges must be a minimum. Finding such a partition is an NP-complete problem. Each subgraph is processed in parallel on a different node of the computing cluster. Communication between subgraphs in the nodes happens through *message passing*.

Every vertex belongs to a *master* subgraph which stores the updated properties of that vertex. At the boundary of the subgraphs, edges span subgraphs. The edge $e : u \rightarrow v$ of a subgraph G_i is called a *remote edge* if it connects to a different subgraph G_j with the master subgraphs of the vertices u and v being G_i and G_j respectively. Information propagation via a remote edge results in communication between the nodes storing G_i and G_j. Latency of such a communication is typically much higher than that in in-memory processing on a CPU or GPU. There is good work balance among computing nodes when each subgraph requires similar amount of computation. Graph partitioning methods should ensure work balance and also minimize the amount of communication between subgraphs, by minimizing the number of remote edges. This is a hard problem and all graph partitioning strategies rely on heuristics in trying to achieve this goal.

Many frameworks have adopted *random* partitioning. Here, each vertex of the graph is considered in turn, and is randomly assigned to a node of the computing cluster. This method achieves good balance in the number of vertices per partition, but there is no control on either the number of remote edges or the number of edges in each partition. The two methods of graph partitioning that provide more control on these issues are *vertex-cut* and *edge-cut* methods. In vertex-cut partitioning, each edge $e \in E$ is assigned a random computing node. Vertex-cut partitioning provides good balance in the number of edges per computing node, but results in communication overhead as vertices can be shared between partitions. For example, two edges $e_1 : u \rightarrow v$ and $e_2 : u \rightarrow w$ with a common tail vertex u, can reside in different subgraphs, thereby sharing vertex u. When a vertex is shared between two or more computing nodes, all the updates to the data in vertex u in any one of the computing nodes must be communicated to the other computing nodes that share the vertex u. On the other hand, in *edge-cut* partitioning, a vertex $v \in V$ is assigned a random computing node, and all the edges $e \in E$ with *tail* $(e) = v$ are added to G_i, the subgraph stored at the computing node. This partitioning method can reduce inter-node communication overhead but may result in improper work balance among computing nodes. A vertex-cut partitioning algorithm that produces balanced partitions and also minimizes communication overheads is described in [1].

Figure 1.3 shows one possible *edge-cut* and *vertex-cut* partitioning on a simple graph for $k == 3$, i.e., three subgraphs. In the edge-cut partition, remote edges are 0–3, 2–3, 2–5, and 4–5. Whenever information is required to flow from vertex 2 to vertex 5, it flows along a remote edge and this results in inter-node communication in the computing cluster. In the vertex-cut partition, vertices 3 and 5 are shared vertices.

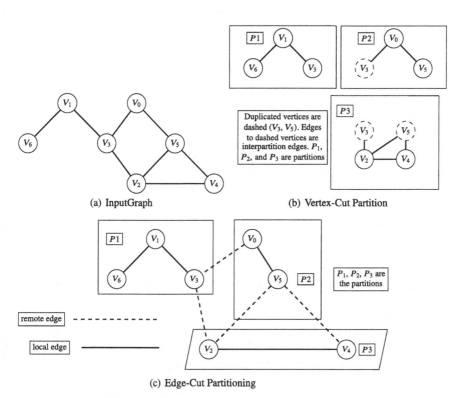

(a) InputGraph

(b) Vertex-Cut Partition

(c) Edge-Cut Partitioning

Fig. 1.3 Graph partitioning: vertex-cut (**b**) and edge-cut (**c**) for graph in (**a**) for $k == 3$

Note that vertex 3 is shared among three partitions and whenever information at vertex 3 is updated in any one of the three partitions, it must be communicated to vertex 3 in the other two partitions.

1.5 Heterogeneous Distributed Systems

A distributed system consists of many computers, geographically distributed over many locations, with remote communications. Most of the time, components in a distributed system work independently, in parallel, and with little interaction between themselves. However, computers in a highly parallel machine have much more coordination than the computers in a distributed system. The computers in a highly parallel machine are physically located very close to each other, and are connected by an extremely fast interconnection network. They share neither a global state nor a global clock, but work in unison to solve a single problem, such as a distributed version of a shortest path algorithm. The communication between such computers is usually using the Message Passing Interface (MPI) and parallelism within each computer is harnessed using threads. Each computer may be of a different type and hence the name heterogeneous. In this book, we refer to such highly parallel heterogeneous machines as *Heterogeneous Distributed Systems*. They are also called Heterogeneous Computing Clusters in literature.

Computing nodes of a distributed system can be differentiated based on their hardware architecture. A multi-core CPU follows Multiple Instruction Multiple Data (MIMD) architecture and each core has a separate program counter. A multi-core CPU executes different instructions at the same time in various cores using separate program counters. A Graphical Processing Unit (GPU) follows Single Instruction Multiple Thread (SIMT) architecture. In the SIMT architecture of a GPU, a set of threads called a *warp*, has a single program counter and executes the same instruction at any point of time. The output of an instruction for a particular thread is committed to memory only if that thread satisfies the condition under which the instruction should execute.

The current generation GPUs have thousands of streaming processors (SPs) which are divided into multiple streaming multi-processors (SMs). The number of SPs in each SM is a multiple of 32, which is the *warp* size. Volatile random access memory available in a GPU is much lesser than that in a multicore CPU. The GPU cores run at a lower frequency compared to a multi-core CPU, but GPUs are more powerful due to the large number of computing cores that they possess, leading to improved throughput.

Figure 1.4 shows a schematic for an Nvidia-Tesla K40 GPU. It consists of 2880 SPs. The SPs are clustered into 15 SMs with each SM having 192 SPs. A K40 GPU can have upto 12 GB volatile memory shared by all SMs. A K40 GPU runs at a frequency of 745 MHz. Each SM has a private memory which is shared among 192 cores (SPs) of the SM. Programs for the Nvidia-GPUs are written using CUDA or OpenCL. A function executed on a GPU is called a *kernel*. If the threads in a

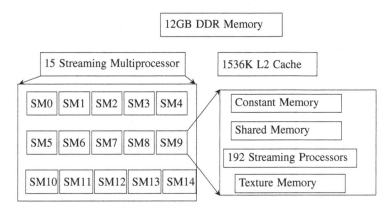

Fig. 1.4 Nvidia-K40 GPU architecture

warp of a GPU kernel follow different execution paths based on runtime values of conditions in the program, it will result in performance degradation. This is called as *warp divergence*. A program written for a GPU should have very few divergent conditional blocks to reduce warp divergence.

Due to the differences in the hardware architecture and programming aspects of CPU and GPU, separate programs are required to be written for the same algorithm. A programmer should know the details of the hardware in order to obtain the best performance. This is a challenging task for a *naïve* programmer with expertise in algorithm design but not with tuning a program for a specific hardware.

1.6   Programming Libraries and APIs

Parallel and distributed systems can be programmed using high level programming libraries. MPI is a *defacto* standard for distributed processing, and OpenMP is popular for multi-core processing. CUDA and OpenCL are widely used frameworks for GPU programming.

1.6.1   OpenMP

The OpenMP [2] application programming interface (API) is used to parallelize loops on a multi-core CPU with shared memory. It consists of compiler pragmas,a runtime library, and environment variables that influence run-time behavior. On a multi-core CPU, each core runs different instructions with different data values at the same time. A loop can be parallelized if there is no data dependency between two or more iterations of a loop. The data dependencies of interest are true-dependency

(Read-After-Write), anti-dependency (Write-After-Read), and output-dependency (Write-After-Write). However, ensuring the absence of data dependencies in a loop is the responsibility of the programmer. When a loop is parallelized using the *omp parallel* pragma, multiple threads are created to execute the iterations concurrently. The compiler handles insertion of appropriate library routine calls into the code. Nested parallelism can be enabled or disabled by setting the OMP_NESTED environment variable. Nested parallelism creates a large number of threads However, if the hardware does not have enough resources (number of cores and memory) to support a large number of threads, then the programmer may decide to disable nested parallelism.

Algorithm 1.1: Parallel Matrix Addition Using `OpenMP` on Multi-Core CPU

```
 1  #include<stdio.h>
 2  #include<stdlib.h>
 3  #include<omp.h>
 4  #define n 8
 5  #define m 8
 6  void readValues(int arr[n][m]) {
 7          for (int i=0; i<n; i++)
 8              for (int j=0; j<m; j++) scanf("%d", &arr[i][j]);
 9  }
10  void printValues(int arr[n][m]) {
11          for (int i=0; i<n; i++){
12              for (int j=0; j<m; j++) printf("%10d", arr[i][j]);
13              printf("\n");
14          }
15  }
16  void main(int argc, char *argv[]) {
17      int rows=n, cols=m;
18      int arr1[rows][cols], arr2[rows][cols], arr3[rows][cols];
19      readValues(arr1); //read first matrix
20      readValues(arr2); //read second matrix
21      #pragma omp parallel for num_threads(3)
22      for (int i=0; i<rows; i++) {
23          for (int j=0; j<cols; j++) {
24              arr3[i][j] = arr1[i][j] + arr2[i][j];
25          }
26      }
27      printValues(arr3);
28  }
```

Program 1.1 shows parallel matrix addition using OpenMP. The pragma in Line 21 creates a chunk of 3 independently running threads. The `for` loop (Lines 22–25) computes the sum of the array elements `arr1[i][j]` and `arr2[i][j]`, and stores it in `arr3[i][j]`. Since the number of threads (3) is not a multiple of the number of iterations of the *outer loop* (8), the assignment of number of iterations to threads is $3, 3, 2$. That means, thread 0 and 1 execute

3 iterations each, and thread 2 executes 2 iterations. The three threads end up computing the sums in 3, 3, and 2 rows of the variable `arr3`, respectively.

1.6.2 CUDA

Algorithm 1.2: Matrix Addition Using CUDA on Nvidia-GPU

```
1  #include <iostream>
2  #include <cuda.h>
3  __global__ void MatrixAdd(int *A, int *B, int *C) {
4      int i = blockIdx.x * blockDim.x + threadIdx.x;
5      C[i] = A[i] + B[i];
6  }
7  void readValues(int *arr, int n, int m) {
8          for (int i=0; i<n; i++)
9                  for (int j=0; j<m; j++) arr[i*m+j]=i+j;
10 }
11 int main() {
12     int rows=1024, cols=1024;
13     int N = rows*cols;
14     // allocate arrays on host (CPU)
15     int *h_arr1 = (int *)malloc(sizeof(int)*N);
16     int *h_arr2 = (int *)malloc(sizeof(int)*N);
17     int *h_arr3 = (int *)malloc(sizeof(int)*N);
18     int *d_arr1, *d_arr2, *d_arr3;// allocate arrays on device (GPU)
19     cudaMalloc((void **)&d_arr1, N*sizeof(int));
20     cudaMalloc((void **)&d_arr2, N*sizeof(int));
21     cudaMalloc((void **)&d_arr3, N*sizeof(int));
22     // Read the input arrays on the host side
23     readValues(h_arr1, rows, cols); //read first matrix
24     readValues(h_arr2, rows, cols); //read second matrix
25     // Copy input arrays from host to device
26     cudaMemcpy(d_arr1, h_arr1, N*sizeof(int), cudaMemcpyHostToDevice);
27     cudaMemcpy(d_arr2, h_arr2, N*sizeof(int), cudaMemcpyHostToDevice);
28     // Make a kernel call for MatrixAdd
29     MatrixAdd<<<rows,cols>>>(d_arr1, d_arr2, d_arr3); //compute on device
30     // Make host wait for the device to finish
31     cudaDeviceSynchronize();
32     // Copy result arrays from device to host
33     cudaMemcpy(h_arr3, d_arr3, N*sizeof(int), cudaMemcpyDeviceToHost);
34 }
```

CUDA[1] [3] is a parallel processing platform and application programming interface (API) model developed by Nvidia for programming Nvidia-GPUs. Discrete

---

[1]CUDA is no longer an acronym.

GPUs have separate memories from the CPU they are associated with. CUDA provides library functions for allocating memory on GPU, copying data to and from the GPU, making kernel calls, etc. It extends C++ language with additional keywords. The CPU is called as the *host* and GPU is called as the *device*. The keyword __global__ precedes the definition of functions which are executed in parallel on the GPU (*device*), and called from the CPU (*host*). Such a function is called a CUDA *kernel*. A CUDA *kernel* is launched with X number of thread blocks and N number of threads per block, where maximum values of X and N are dependent on the specific GPU device. CUDA kernel calls are asynchronous and the *host* continues execution after a kernel call. A *barrier* synchronization function *cudaDeviceSynchronize()* forces the *host* to wait for the *device* to complete its operation.

Program 1.2 performs matrix addition on GPU using the CUDA *kernel* Matrix-Add (Lines 3–6). In the main program, first the arrays are allocated on CPU and GPU using *malloc()* and *cudaMalloc()* functions respectively. This is followed by reading of the two input arrays. Even though the matrices are two dimensional, they are being stored into one dimensional arrays in row-major order. Single dimensional arrays are easier to copy from the host to the device and vice-versa. The array elements are then copied from CPU (*host*) to GPU (*device*) using the *cudaMemcpy()* function. The last argument in the *cudaMemcpy()* function specifies the direction of data transfer as *host* to *device* (see Lines 26, 27).

Then the CUDA *kernel* MatrixAdd is called, which performs the operation $d\_arr3[i][j] = d\_arr1[i][j] + d\_arr2[i][j], such that 0 \leq i, j < 1024$, but on corresponding one dimensional arrays. Since a separate thread is created for computing each element of d_arr3, each thread must know which element of d_arr3 it must compute and which elements of d_arr1 and d_arr2 must be used to compute it. This index is computed in the variable i (Line 4) using the variables *blockIdx.x* and *blockDim.x* which store the *block number* and the number of blocks in the CUDA kernel respectively. The variable *threadIdx.x* stores thread-id within the block. In this example, computation of i is simple because a separate thread is created for computing each element of d_arr3. Finally, the result stored in the array d_arr3 is copied from *device* to *host* (see Line 33).

1.6.3 OpenCL

Open Computing Language (OpenCL) is a language and framework for programming heterogeneous systems that include CPU, GPU, FPGA, DSP and other accelerators. OpenCL has a C-like syntax. A function executed on a device is called a *kernel* in OpenCL also. The programming language for writing *kernels* is called OpenCL C. OpenCL kernels can be compiled offline or at *runtime*. However, compilation of kernels at runtime involves some latency and caching of compiled code is employed in order to minimize this latency.

Program 1.3 shows a complete OpenCL program to compute the square of integers in an array. The host issues the command to acquire the ID of the device on which computation will occur (Line 28) and specifies the device type (CPU, GPU, ACCELERATOR or ALL) in the command. A mere change in the device type specification makes the application portable to any other type of device. Context is an environment that contains one or more devices and their memories of the same platform and is used by the OpenCL runtime (created at Line 29). Each device is associated with a command queue which is used to enqueue commands to that device such as, write buffer, read buffer and launch kernel. The command queue is created at Line 30 in this example.

Memories for input vector and output vector are created both on the host and on the device. Input memory on the device is kept READ_ONLY (see Line 31) and output memory is kept WRITE_ONLY (see Line 32) because work items read the input data and write the output data in the kernel. Input data is copied from host to device by issuing a command to the command queue (see Line 33).

The kernels are written in a separate file (.cl file) which is read by the host (see Line 22) to create a *program* from it (see Line 34). A *program* is a collection of all the kernels, compiled either online or offline. In the example, the kernel is compiled online (see Line 35). The kernel is created (see Line 36) and launched with size equal to vector length (Line 41). The result is copied back from the device to the host (see Line 43).

1.6.4   Thrust

Thrust [4] is a CUDA library similar to C++ Standard Template Library, and it eases programming parallel applications. Thrust library has data parallel primitives such as sort, reduce, scan, etc. These primitives can be used to implement complex parallel applications. The thrust library avoids using *cudaMemcpy()* and *cudaMalloc()* for explicit memory transfer operation and memory allocation respectively. This makes programming GPUs simpler. The program shown in Algorithm 1.4 is for matrix addition using the thrust library. The program uses the *plus* transformation for each element in the arrays *h_arr1* and *h_arr2*. The result is stored in the vector *d_arr3*.

1.6.5   MPI

The Message Passing Interface (MPI) [5] is a message passing library standard put forth by the MPI Forum. There are several implementations of MPI libraries available in public which have been built according to the specifications of MPI. The MPI library is used for communication between the nodes of a distributed system with each computing node having private memory. It primarily focuses

Algorithm 1.3: OpenCL: Kernel and C Program

```
1  __kernel void findsqaure(__global int *A, __global int *B) {
2      int id = get_global_id(0);
3      B[id] = A[id] *A[id];
4  }
5  //C Program
6  #include <stdio.h>
7  #include <stdlib.h>
8  #include <inttypes.h>
9  #include <errno.h>
10 #include <string.h>
11 #include <CL/cl.h>
12 #define SRC_SIZE (10000)
13 #define val_size 1000
14 char val[val_size];
15 char *srcpath= "./square.cl";
16 int main(void) {
17     // Create the two input vectors
18     int *X = (int *)malloc(sizeof(int)*1024);
19     for(int i = 0; i < 1024; i++) X[i]=i;
20     FILE *fp = fopen(srcpath, "r");
21     char *src_str = (char *)malloc(SRC_SIZE);
22     size_t src_size= fread( src_str, 1, SRC_SIZE, fp);
23     cl_platform_id platform_id = NULL;
24     cl_device_id device_id = NULL;
25     cl_uint ret_num_devices;
26     cl_uint ret_num_platforms;
27     clGetPlatformIDs(1, &platform_id, &ret_num_platforms);
28     clGetDeviceIDs(
       platform_id,CL_DEVICE_TYPE_GPU,1,&device_id,&ret_num_devices);
29     cl_context context = clCreateContext( 0, 1, &device_id,0,0,0);
30     cl_command_queue cmd_queue = clCreateCommandQueue(context, device_id, 0,0);
31     cl_mem X_dev =clCreateBuffer(context, CL_MEM_READ_ONLY, 1024 * sizeof(int),
       0,0);
32     cl_mem Z_dev =clCreateBuffer(context, CL_MEM_WRITE_ONLY, 1024 *
       sizeof(int), 0,0);
33     clEnqueueWriteBuffer(cmd_queue, X_dev, CL_TRUE, 0,1024 * sizeof(int), X, 0, 0,
       0);
34     cl_program prg =clCreateProgramWithSource(context,1,(const
       char**)&src_str,&src_size,0);
35     cl_int ret = clBuildProgram(prg, 1, &device_id, 0, 0,0);
36     cl_kernel kernel = clCreateKernel(prg,"findsquare",0);
37     clSetKernelArg(kernel, 0, sizeof(cl_mem), &X_dev);
38     clSetKernelArg(kernel, 1, sizeof(cl_mem), &Z_dev);
39     size_t g_size = 1024; // Process the whole list
40     size_t l_size = 64; // Process in groups of size 64
41     clEnqueueNDRangeKernel(cmd_queue, kernel, 1, 0, &g_size, &l_size, 0, 0, 0);
42     int *Z = malloc(sizeof(int)*1024);
43     clEnqueueReadBuffer(cmd_queue, Z_dev, CL_TRUE, 0,1024 * sizeof(int), Z, 0, 0,0);
44     for(int i = 0; i < 10; i++) printf("log(%d) = %d", X[i], Z[i]);
45     return 1;
46 }
```

Algorithm 1.4: Matrix Addition Using Thrust Library

```
 1  #include <thrust/sort.h>
 2  #include <thrust/device_vector.h>
 3  #include <thrust/host_vector.h>
 4  #include <thrust/transform.h>
 5  #include <thrust/sequence.h>
 6  #include <thrust/copy.h>
 7  void readValues(int *arr, int n, int m) {
 8      for (int i=0; i<n; i++)
 9          for (int j=0; j<m; j++) arr[i*m+j]=i+j;
10  }
11  main(int argc, char *argv[] ) {
12      int rows=1024, cols=1024;
13      int N = rows*cols;
14      // allocate arrays on host (CPU)
15      int *h_arr1 = (int *)malloc(sizeof(int)*N);
16      int *h_arr2 = (int *)malloc(sizeof(int)*N);
17      thrust::host_vector<int> h_arr3(N);
18      readValues(h_arr1,rows,cols);
19      readValues(h_arr2,rows,cols);
20      thrust::device_vector<int> d_arr1(h_arr1,h_arr1+N);
21      thrust::device_vector<int> d_arr2(h_arr2,h_arr2+N);
22      thrust::device_vector<int> d_arr3(N);
23      thrust::transform(d_arr1.begin(), d_arr1.end(), d_arr2.begin(), d_arr3.begin(),
           thrust::plus<int>() );
24      thrust::copy(d_arr3.begin(), d_arr3.end(), h_arr3.begin());
25  }
```

---

on the message-passing parallel programming model. It defines the notion of a *communicator* (MPI_COMM_WORLD). A communicator defines a group of processes that may communicate with one another. In this group, each process is assigned a unique *rank*, and the processes communicate with one another using their ranks. The communication between processes is programmed using *message passing*, and is built upon *send* and *receive* operations. Communication involving one sender and one receiver is called *point-to-point* communication.

The MPI APIs for *send()* and *receive()* operations are *MPI_Send()* and *MPI_Recv()* respectively. The API functions take as arguments data, its type, and its size to be sent or received, *message-id*, and *rank* to identify the process on the remote node. The *MPI_Recv()* function blocks until data is received from the process with the *rank* given as the argument. Asynchronous communication between nodes is possible with the *MPI_Isend()* and *MPI_Irecv()* function calls.

Program 1.5 illustrates a sample MPI code. In the program, the process with *rank* zero (Lines 7–10) sends an integer value of -1 to the process with *rank* one (see Line 9). The process with *rank* one (Lines 11–14) blocks until the value is received from process with *rank* zero (see Line 12). Once the value is received by the process with *rank* one, the received rank and value are printed (see Line 13).

Algorithm 1.5: MPI Send and Receive Sample Program

```
 1  #include<mpi.h>
 2  int main(int argc,char *argv[]) {
 3      int rank,size; int number=0;
 4      MPI_Init(&argc,&argv);
 5      MPI_Comm_rank(MPI_COMM_WORLD, &rank);
 6      MPI_Comm_size(MPI_COMM_WORLD, &size);
 7      if( rank == 0 ){
 8          number = -1;
 9          MPI_Send(&number, 1, MPI_INT, 1, 0, MPI_COMM_WORLD);
10      }
11      if( rank == 1 ){
12          MPI_Recv(&number, 1, MPI_INT, 0, 0,
                    MPI_COMM_WORLD,MPI_STATUS_IGNORE);
13          printf("Process %d received integer value %d from process 0", rank, number);
14      }
15      MPI_Finalize();
16  }
```

MPI also supports collective operations, such as Barrier, Broadcast, Scatter, Gather, and reduction (add, multiply, min, max, etc.) on a group of processes:

1. The function `MPI_Barrier(comm)` blocks the caller until all the processes have reached this routine. Then, they are all free to proceed with their own executions. The parameter `comm` is the group of processes to be synchronized. This routine is used to synchronize processes to ensure that all of them reach a predetermined point in computation before proceeding further.

2. `MPI_Bcast(&buffer, count, datatype, root, comm)` is the broadcast routine that broadcasts `count` number of data of type `datatype` from the variable `buffer` in the process with rank as `root`, to all the processes of the group `comm`.

3. The routine `MPI_Scatter(&sendbuf, sendcnt, sendtype, &recvbuf, recvcnt, recvtype, root, comm)` is meant to distribute distinct messages (`sendcnt` in number), from the buffer `sendbuf`, and from a single source process `root`, to each process in the group `comm`. The messages will be received in `recvbuf` of each process. `MPI_Gather(...)` is the reverse operation of Scatter. It collects distinct messages from each process in the group in a single destination process.

4. `MPI_Reduce(&sendbuf, &recvbuf, count, datatype, op, root, comm)` applies a reduction operation (`op`) on the data in all the processes in the group `comm`, and places the result in one process `root`. `op` can be one of the many predefined operators mentioned above or can be provided by the programmer. Figures 1.5 and 1.6 show the operation of these four routines.

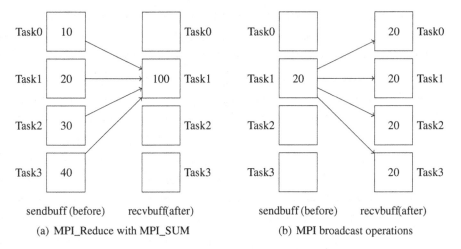

Fig. 1.5   MPI_Reduce (SUM) and MPI_Bcast operation

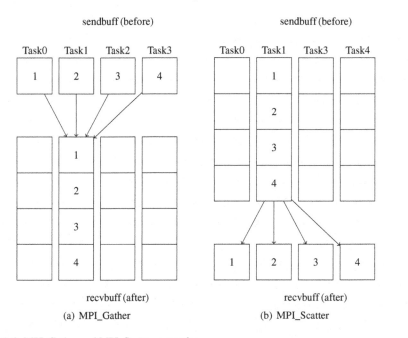

Fig. 1.6   MPI_Gather and MPI_Scatter examples

MPI library is quite vast and [5] contains a good description of the routines provided in it.

1.7 Graph Analytics in Real-World Applications

Graphs can model several real-world phenomenon. For instance, we can view various webpages as nodes and hyperlinks as directed edges to represent the world wide web as a graph. Such a modeling can help perform various graph computations on the web. For instance, the pagerank algorithm is a popular graph algorithm, which is used to rank the webpages. Alternatively, the web-graph can be used to find clusters of webpages which link one another. This can help in categorizing the webpages into various topics.

Graphs can also naturally model road networks. The Single Source Shortest Path (SSSP) computation is often used to find the shortest distance between two endpoints in a map. The Breadth First Search (BFS) algorithm is used in social network analysis to find the distance (hop) between two persons. The triangle counting algorithm is used in community detection. The connected component algorithm is used to cluster a graph into different subgraphs.

1.8 Graph Analytics Challenges

Graphs pose challenges for parallelization due to their inherent input-centric nature. We discuss those challenges below.

1.8.1 Classification of Graphs

Graphs and their properties have been discussed in Sect. 1.2. Some properties such as diameter and variance in degree are very important to graph algorithms and their execution on different types of hardware. The difference in these properties make a program perform poorly or very well for input graphs of different types (road, random, R-MAT). Table 1.5 shows a comparison of different graph types based on the properties.

In a *road network graph*, a vertex represents a junction of two or more roads and an edge represents a road connecting two junctions. Road network graphs have a very high diameter and a low variance in degree. A *random* graph is created from a set of vertices by randomly adding edges between the vertices. Erdös and

Table 1.5 Graph classes and their properties

Graph type	Diameter	Degree-variance
Random	Low	Low
R-MAT	Low	High
Social N/W	Low	High
Road N/W	High	Low

Rényi [6] model assigns equal probability to all graphs with V vertices and E edges.

Real world graphs of social networks such as Twitter can have multiple edges between the same pair of vertices. Such graphs could be *unipartite* like people in a closed community group, *bipartite* like in a *movie-actor* or *author-publication* database and possibly multipartite. Such graphs follow the *power-law* degree distribution with very few vertices having very high degree (indegree or outdegree) [7]. As an example, in a Twitter network, a node corresponding to a celebrity who has a large following will have a very high degree. Social network graphs have a very low diameter unlike a road network. This is called as the *small-world phenomenon* [8]. R-MAT [9] graphs follow the power-law distribution and can be created manually. The R-MAT graph generator recursively subdivides the adjacency matrix of a graph into four equal-sized partitions, and distributes edges to these partitions with unequal probabilities a, b, c, d such that (a + b + c + d) = 1.

A hypergraph G(V, E) is a graph where V is the set of vertices, and E is a set of non-empty subsets of V called hyper-edges or edges. A *k*-uniform hypergraph will have all edges *e* in E having *v* vertices. A two uniform hypergraph is an ordinary graph. Hypergraphs have applications in game theory, data mining, etc.

1.8.2   Irregular Computation

In irregular computations, data-access or control patterns are unpredictable at compile time. This happens due to input-dependent access patterns. In contrast, in regular computations, the access patterns are known at compile time. For instance, in dense matrix-matrix computation, without knowing the values inside the two matrices (which are available at runtime), a compiler can identify which elements get accessed (row elements of the first matrix along with the column elements of the second matrix). Thus, the access pattern is known from the program without knowing the input. In contrast, in graph algorithms, without knowing the input graph, it is unclear which vertex would be connected to which other vertices. Thus, the edges along which, vertex attributes would be propagated, is not clear at compile time. Such an irregularity poses challenges in optimizations, as well as in automatic parallelization. Compilers need to make conservative assumptions about the input, which forbids parallelization (e.g., in the absence of any other information, a compiler may assume that any edge is probable in the graph, leading to pessimistic assumptions with reduced parallelism).

Therefore, irregular computations demand dynamic processing, wherein, parallelism decisions are deferred until runtime when the input graph is available. This adds runtime overhead to processing. In addition, several real world graphs have arbitrary connectivity leading to load imbalance. As discussed, several graphs such as social networks have skewed degree distribution. If two threads concurrently operate on two vertices, it is possible that the thread operating on a low-degree vertex would have to wait for the other thread operating on a high-degree vertex, without performing useful work.

Graphs also pose synchronization challenges. Unlike traditional regular programs, wherein different array elements are processed independently, graph algorithms may involve shared data being read and written to by concurrent threads. For instance, two threads may write to distance of the same node in shortest paths computation. Such a *racy* operation demands usage of synchronization constructs (such as locks, barriers, atomics, etc.) leading to further slow-down of the overall algorithm.

1.8.3 Heterogeneous Hardware

A heterogeneous hardware system includes one or more multi-core CPUs and GPUs. A multi-GPU machine has more than one GPU and one multi-core CPU. When programs are executed on a multi-core CPU, the graph storage format, size of the graph and graph properties listed in Table 1.5 determine how fast the program will complete its execution. Particular graph types require implementations using particular data structures in order to be efficient. An excellent example is the Δ-stepping algorithm [10] and its implementation on a multi-core processor. This algorithm uses a dynamic *collection* called *bucket* and has proved to be very efficient for road network graphs which have a very high *diameter*. However, implementations of the Δ-stepping algorithm on GPUs are generally not as efficient as on multi-cores.

1.8.4 Distributed Execution

A distributed system with multi-core CPUs, GPUs, and multi-GPU systems is a platform where graph analytics can be performed on very large graphs. Performance of distributed execution depends on graph partitioning heuristics and communication overhead, in addition to the graph structure. If the GPUs of a multi-GPU machine are on same bus, the peer-access feature of CUDA[2] can be used to reduce the communication overhead. If communication happens over a network via ethernet or WiFi, then communication time depends on the speed of the network.

---

[2]For Nvidia-GPUs only.

Algorithm 1.6: Pseudo-Code for Pagerank Algorithm

```
1  pagerank(Point p, Graph graph) {
2      double val = 0.0;
3      foreach( t in p.innbrs ) in parallel {
4          if( t.outdegree!=0 ){
5              val += t.pr / t.outDegree;
6          }
7      }
8      p.pr = ADD(val * d, (1 - d) / graph.npoints, -1);
9  }
10 main() {
11     ......
12     foreach( t in graph.vertices ) in parallel {
13         t.pr = 1 / graph.npoints;
14     }
15     int cnt = 0;
16     while( 1 ){
17         foreach( t in graph.vertices ) in parallel {
18             pagerank(t, graph);
19         }
20         if ( ITERATIONS == cnt ){
21             break;
22         }
23         ++cnt;
24     }
25     .....
26 }
```

1.8.5 *Atomic Operations and Synchronization*

Algorithm 1.6 shows the pseudo-code for computing the pagerank of each vertex in a graph.[3] After reading the graph the pagerank of each vertex is initialized to $1 \div |V|$ (Line 12) in a parallel loop. Then, in the `while` loop the pagerank value (pr) of each vertex is updated using a call to the function *pagerank* (Lines 1–9). The `foreach..in parallel` statements in Lines 17, 12, and 3 are parallel loop statements, in which each iteration of the loop is executed in parallel.

This algorithm updates the pagerank value of a vertex p by *pulling* values of the vertices $t \in V$ such that there is an edge $t \rightarrow p \in E$ of the graph (Line 3). In such an implementation, the variable *val* can be updated without an atomic operation. But if the pagerank value is computed by a *push* model, where for all edges $p \rightarrow t$ the pagerank value of the vertex p is added to the pagerank value of vertex t, then the `forall` loop should add *val* using an atomic operation. This is due to the irregular

---

[3]See Sect. 2.3.1 of Chap. 2 for a detailed explanation of the pagerank algorithm.

nature of the graph algorithm. Push and Pull versions of parallel *pagerank* algorithm are discussed in Sect. 3.2.5.

Synchronous execution involves a barrier after the *forall* call to the function *pagerank()* (Line 17). This ensures that the program waits for the parallel call to the function *pagerank()* to finish. But if the algorithm is executed asynchronously, there will be no barrier. Such a feature is useful when the computation happens using CPU and GPU. The *host* (CPU) calls the *device* (GPU) function pagerank, and the host continues execution without waiting for the device to finish the execution of the *pagerank()* function.

1.8.6 Challenges in Programming

Programming graph analytics targeting a distributed system with CPUs and GPUs is very challenging. The programmer must handle graph partitioning, parallel computation, dynamic memory management, and communication between nodes. Multi-core CPUs and GPUs follow MIMD and SIMT architectures respectively. The graph storage format should maximize cache locality and coalesced access to obtain high throughput. Thread management for CPUs and GPUs is very challenging. Programs written in native languages such as C or CUDA with libraries such as OpenMP and MPI will be very large. Such programs will be difficult to understand and modify, and are error-prone. The complexity of programming graph analytics for distributed systems can be reduced by high level programming abstractions.

1.9 Programming Abstractions for Graph Analytics

Programming abstractions are used in various domains. Well known examples are VHDL for hardware design, HTML for web programming, and MATLAB for scientific computations. Such domain-specific frameworks or languages make the job of coding and debugging much easier than in the usual programming languages. More specifically, abstractions for heterogeneous hardware, graphs and distributed computation will help in programming graph analytics applications. This increases productivity and reliability.

1.9.1 Frameworks

Frameworks make programming an algorithm much easier than with a library. Frameworks provide a generic API [11, 12] for the implementation of the algorithm. Programmer implements different parts of the algorithm (computation, communication, aggregation) according to requirements of specific functions of the API. The flow computation is restricted by the framework due to such API, but it provides modularity and a better programming style, apart from makeing it comfortable to program algorithms. Many graph analytics frameworks have been proposed in the past which differ in multiple dimensions. Frameworks which work only on single multi-core CPU machines are not scalable with graph size. Large graphs that do not fit into the memory of a single machine cannot be used with such frameworks. Some frameworks support changes in graph topology during computation, some promise better performance using *speculative* execution, and some others support efficient worklist based implementations. A distributed system is required for processing large graphs. Different execution models have been proposed in the past for distributed execution on a CPU cluster. These include the Bulk Synchronous Parallel (BSP) model of execution, the *Gather-Apply-Scatter* execution model, and the asynchronous execution model. This section provides an overview of such models. Details of these models along with examples are covered in later chapters.

1.9.1.1   Bulk Synchronous Parallel (BSP) Model

In the BSP model of programming [13], input data is partitioned on the multiple compute nodes. Each compute node computes the algorithm locally, and subsequently, communicates with other nodes at the end of computation. Programs in the BSP model are written as a sequence of iterations called *supersteps*. Each superstep consists of the following three phases:

1. Computation: Each compute node performs local computation independently on its partitioned data and is unaware of execution on other nodes. Nodes may produce data that needs to be sent to other nodes in the communication phase.
2. Communication: In the communication phase, data is exchanged as requested in the previous computation phase.
3. Synchronization: There is an implicit barrier at the end of communication phase. Each node waits for data which was transferred in communication phase to be available.

The BSP model is used to simplify the programming in a distributed environment. In the context of graph processing, each vertex behaves as a computing node, and vertices communicate among each other through edges. Pregel is an example of such a framework [14] based on the BSP model. The *MapReduce()* framework on the Hadoop distributed file system (HDFS) for storing data is adaptable for graph analytics. Giraph [15] is an example of such a framework that combines the benefits of Pregel and HDFS.

1.9.1.2   Asynchronous Execution Model

In contrast to the BSP model (which is synchronous), asynchronous systems update parameters using the most recent parameter values as input. Frameworks based on BSP may incur penalties when the compute nodes of the distributed system have varying speeds and/or the communication speeds between all nodes are not uniform. This is because synchronization waits for the *slowest* machine to finish. Imbalance in finishing times of nodes may also be due to the varying amount of computation in the nodes. For example, consider a power-law based graph that is randomly partitioned. Some nodes may have vertices which have a very large degree, thereby requiring more computation time than other nodes.

In such situations, resorting to asynchrony, where nodes do not always wait for all others to finish, may make the distributed computation execute faster. However, asynchrony comes with its own dangers of non-determinism which could yield inconsistent results. This requires a careful provision for consistency control features in the framework. Asynchronous execution systems also require work lists of active elements (vertices in case of graphs) that are scheduled appropriately. An example of a computation at a node, written in a framework such as Graphlab [16] or Distributed Graphlab [17] is shown in Algorithm 1.7.

Algorithm 1.7: Execution with Worklist

---

1 **Input:** Data Graph G = (V, E, D)
2 **Input:** Initial vertex worklist T = $\{v_1, v_2, \ldots\}$
3 **Output:** Modified Data Graph G = (V, E, D')
4 **while(** $(T \neq \phi)$ **){**
5 $\quad$ v $\leftarrow$ GetNext(T)
6 $\quad$ (T', S_v) $\leftarrow$ update(v, S_v)
7 $\quad$ T $\leftarrow$ T $\cup$ T'
8 **}**

1.9.1.3   Gather-Apply-Scatter (GAS) Model

The *Gather-Apply-Scatter* model of execution consists of a super step which is divided into three *minor* steps named *gather*, *apply* and *scatter*. The changes made to the graph properties are *committed* after each minor step. The commit process requires transfer of vertex and/or edge property values between subgraphs stored in the nodes of a computer cluster. In each superstep, a set of *active* vertices are processed. At the end of a superstep, new active vertices are created. The computation stops if no active vertices are created after a superstep, which denotes reaching the *fixpoint* of the computation. Initiation of execution is with a set of algorithm-specific active points (e.g., the source node in the SSSP algorithm). The

Gather-Apply-Scatter model supports both the BSP model and the asynchronous model of execution.

Algorithm 1.8: Gather-Apply-Scatter-Vertex Model

---
1 **interface** GASVertexProgram(u) {
2 // Run gather on neighbours(u)
3 gather(D_u, D (u,v), D_v) $\rightarrow$ Accum
4 sum(Accum left, Accum right) $\rightarrow$ Accum
5 apply(D_u, Accum) $\rightarrow D_u^{new}$
6 // Run scatter on neighbours(u)
7 scatter(D_u^{new}, D(u,v), D_v) $\rightarrow$ ($D_{(u,v)}^{new}$, Accum)
8 }

Algorithm 1.8 shows the pseudo code of the execution model. In the gather phase an active vertex invokes the algorithm-specific gather function on adjacent vertices of vertex *u*. The result is computed using the *sum* function. The sum function of an algorithm should be commutative and associative. The computed result is then scattered using the edges connected to the vertex *u*. Gonzalez et al. [11] implements such an abstraction.

Graph analytics is efficient on GPUs. Implementations of graph algorithms written in languages such as C++/CUDA/OpenCL are efficient, but involve many details of hardware, thread management, and communication. Many frameworks have been developed over the last decade to support graph analytics on GPUs and make programming easier. Such frameworks usually support multiple-GPUs but only on a single machine. If the GPUs are connected with a *peer-access* capability, the communication overhead between GPUs on a multi-GPU machine is very low. Otherwise, it is very high. The communication cost between GPUs on different nodes of a cluster is very high compared to that betweeen CPUs. Therefore, proper graph partitioning assumes a great significance on such systems. LonestarGPU [18] and IrGL [19] supports *cautious morph* graph algorithms (a subset of dynamic graph algorithms) on GPUs. Totem [20] and Gluon [21] are graph analytics frameworks for multi-GPU machines and GPU cluster respectively.

1.9.2   Domain Specific Languages

Graph analytics frameworks lack support for semantic checks of programs. They also lack higher level of abstractions as compared to a domain specific language (DSL). It is relatively more difficult to program algorithms using frameworks than using graph DSLs. In a graph DSL, elementary data items such as *vertex*, *edge* and *graph* are provided as data types with semantics for operations on each data type. Further, DSLs comes with parallel and synchronization constructs, and also data types such as *Collection* and *Set* which are necessary to implement even elementary algorithms. The syntax and semantic violations are caught by the DSL compiler. All this makes programming graph analytics easier and less error prone, thereby increasing productivity.

Several DSLs for graph analytics on multi-core CPUs have been proposed in the past. Some of the Graph DSLs such as Green-Marl [22] and Elixir [23] support only static graphs[4] with mutable *edge* and *vertex* properties. Green-Marl has been extended with support for CPU clusters [24]. Lighthouse [25] extended Green-Marl for Nvidia-GPUs. Graph DSLs targeting single machines with a multicore-CPU, multi-GPUs, and distributed systems with multi-core and multi-GPU configurations have been reported in Gluon [21], Falcon [26], and DH-Falcon [27].

Algorithm 1.9: Bellman–Ford SSSP Code in Falcon

```
 1  int changed=0;
 2  relaxgraph(Point p, Graph graph) {
 3      foreach( t In p.outnbrs ){
 4          |  MIN(t.dist, p.dist + graph.getWeight(p, t), changed);
 5      }
 6  }
 7  SSSP(char *name) {
 8      Graph hgraph;
 9      hgraph.addPointProperty(dist, int);
10      hgraph.read(name);
11      foreach(t In hgraph.points) t.dist = 1234567890;
12      hgraph.points[0].dist = 0;
13      while( 1 ){
14          changed = 0;
15          foreach(t In hgraph.points ) relaxgraph(t, hgraph);
16          if(changed == 0) break;
17      }
18      return;
19  }
20  main(int argc,char *argv[]) {
21      |  SSSP(argv[1]);
22  }
```

In order to demonstrate the benefits of DSLs, Algorithm 1.9 shows the code for the single source shortest path algorithm (Bellman–Ford SSSP algorithm) written in the Falcon DSL. While this code is quite compact, the same algorithm implemented in a framework will be lengthier, more complex, and harder to debug. The important function is *SSSP* in which *hgraph* is declared as a *Graph* type variable, a property *dist* is added to its vertices (Line 9), the input graph is read (Line 10), and *dist* of all vertices except the source is initialized to a very large value (source gets zero value). The variable *changed* keeps track of any changes made to the shortest

---

[4]The vertices and edges of the graph remain unchanged.

distance property, *dist*. The while loop exits when no changes in *dist* are occur for any vertex (Line 16). Within the while loop, shortest distance is updated in parallel for each vertex by calling the function *relaxgraph*. This function updates the shortest distance property *dist* of a vertex using an atomic function MIN. It also updates the variable it changed. More details of the SSSP algorithm are provided in the next two chapters.

Chapter 2
Graph Algorithms and Applications

This chapter provides a discussion of various sequential graph algorithms and issues in their implementations. After describing fundamental algorithms like traversals, shortest paths, etc., more specialized algorithms such as betweenness centrality, page rank, etc. follow. The chapter ends with a focused discussion of applications of graph analytics in different domains such as graph mining and graph databases.

2.1 Introduction

Distributed implementation of graph algorithms and their performance in different domains vary depending on:

- the input graph structure,
- the order of processing the vertices and edges of the graph,
- the way graph properties and structure are modified, and
- graph partitioning and communication overhead.

An adjacency matrix data structure is suitable for graph storage when the number of edges in the graph object is close to $|V|^2$, so that memory utilization is high. Adjacency matrix has the advantage of requiring only a constant access time for each edge in the graph object. CSR storage format has a storage overhead of $O(V + E)$[1] and is suitable for sparse graphs. This storage format is suitable for graph analytics covering many application domains.

Automatic and manual parallelization of regular computations with regular data structures (e.g., matrix) has been explored sufficiently and a reasonable understanding of the same has been acquired. Graph analytics generally uses *irregular* data structures such as unbalanced trees, sparse graphs, unstructured grids, sets, etc.

---

[1]Ideally, this should be $O(|V| + |E|)$, but we use this shorter notation when the context is clear.

© Springer Nature Switzerland AG 2020
U. Cheramangalath et al., *Distributed Graph Analytics*,
https://doi.org/10.1007/978-3-030-41886-1_2

Parallelization of irregular computations (e.g., graph analytics) for heterogeneous hardware is of great importance as analytics focuses on unstructured big data. The challenges involved in this parallelization are still being explored.

Irregular applications in different application domains of graphs are solved using parallel graph algorithms. Parallel graph analytics has proved to be efficient on GPUs and multi-core CPUs. Implementations of parallel graph algorithms rely on synchronizations using atomic operations provided by the hardware in order to tackle race conditions arising from irregular access patterns. Typical atomic operations provided in the current hardware systems include *compare-and-swap*, *atomic-min*, *atomic-add* etc. Placement of synchronization instructions at appropriate program point is an important performance concern. Unnecessary synchronizations hamper parallelism and reduce efficiency of parallel programs.

Irregular control paths pose challenges in programming parallel algorithms. The control paths of each thread depend on the value of the condition in a conditional statement (e.g., if, switch, and loops such as for). *Warp divergence* on GPUs arises due to the threads in the same warp following different control paths, and leads to inefficiency and loss of speedup. Irregular communication patterns occur due to execution of a application in a distributed system which works with an input graph with arbitrary connectivity. Once again, minimizing the amount of communication is important and has a bearing on the speedup. Load balancing across compute nodes is also a challenging problem in graph analytics, which is directly affected by graph partitioning and distribution, apart from the algorithm. Algorithms may be vertex-based or edge-based, and graph partitioning must be based on this knowledge.

A few application domains where graph analytics are used are listed below:

- Social network applications such as community detection, cyber-attack and fraud applications. Examples include:

  - Making product recommendations based on group affiliation or similar items.
  - Understanding consensus in social communities and estimating group stability.
  - In fraud analysis, evaluating whether a group of people has simply a few discrete bad behaviors or is acting as a fraud group.

- Transportation network analysis and design.
- Finding the optimal location of new public services for maximum accessibility
- Finite element methods which use multi-dimensional grids as graphs. Typical examples of finite element methods are Delaunay mesh refinement and delaunay triangulation which produce two and three dimensional meshes.
- Compiler optimizations which use data flow analysis techniques that run on control-flow graphs. Graph coloring algorithms are used by compiler code generators which work with interference graphs.
- Machine learning applications such as belief propagation and survey propagation are programmed with a *factor graph* as the data structure. A factor graph is a sparse bipartite graph, and represents the dependencies between variables and factors (parts of a formula).

- Bio-informatics applications for improvement of drug targeting or finding dangerous combinations of possible co-prescribed drugs
- In natural language processing, for text summarization, sentiment analysis, text ranking, etc.

Almost all the applications mentioned above work with large graphs as input which invariably require distributed computation.

Graph processing typically happens in multiple iterations until a fixpoint is reached. In each iteration, only a subset of vertices or edges is processed. These are called as *active elements* for that particular iteration. The *active elements* lead to processing of their neighbours which could be:

- source and destination points of an edge if the *active element* is an edge.
- neighbouring vertices of a vertex if the *active element* is a vertex.

The operations performed by an algorithm on the input graph can modify graph properties, add or delete points and edges. Based on this feature, a graph algorithm is classified as:

- Local computation—the computation changes only the properties of vertices and edges in the graph, and the graph structure is not modified. For instance, finding the number of hops between two persons on a static social network.
- Incremental dynamic computation—the graph structure is changed by adding vertices and/or edges. For instance, finding the possibly reduced number of hops between two persons on a social network where friendship relations keep getting added.
- Decremental dynamic computation—the graph structure is changed by deleting vertices and/or edges. For instance, finding the possibly larger number of hops between two persons on a social network where people leave the network.
- Fully dynamic computation—the graph structure is changed by adding and deleting vertices and/or edges. For instance, maintaining the number of hops between two persons on a social network where friendship relations keep getting added and removed, is a dynamic computation. Similarly, in transportation networks, where the edge weight is stored as time to travel from one junction (Vertex) to another, the edge weight may change depending on the time of the day (high at 9 a.m. and low at 11 p.m.).

We now discuss some of the important elementary graph algorithms in detail.

2.2 Fundamental Graph Algorithms

In this section we discuss elementary graph algorithms which have many applications. For simplicity, explanation of an algorithm considers only the sequential execution. However, parallelization and distributed executions of these algorithms are very important considering the huge graphs on which they operate. They will be discussed in detail in the forthcoming chapters. *The algorithmic notation in this chapter uses* `Point` *and* `Vertex` *interchangeably.*

2.2.1   Breadth-First Search (BFS)

Algorithm 2.1: Breadth-First Search (BFS) Traversal

```
 1  BFS(Vertex src, Graph G) {
 2      foreach( Vertex p In G ){
 3          p.dist = ∞;
 4          p.pred = NULL;
 5      }
 6      src.dist = 0;
 7      while( True ){
 8          changed = False;
 9          foreach( Vertex p In G ){
10              foreach( Vertex t In p.outnbrs ){
11                  if( t.dist >(p.dist + 1) ){
12                      t.dist = p.dist + 1;
13                      t.pred = p;
14                      changed = True;
15                  }
16              }
17          }
18          if (changed == False) break;
19      }
20  }
```

BFS is one of the simplest graph traversal techniques [28]. The algorithm takes as input a directed or undirected graph G(V, E) and a source vertex *src* (see Algorithm 2.1). The traversal explores the edges of G to find all the vertices that are reachable from the source vertex *src* in a level-order fashion. It computes a path with the smallest number of edges to all the vertices reachable from *src*. A few applications of BFS include:

* *Web crawlers* to build the index
* Social networks to find members with a distance k from a person
* GPS to find nearby locations.

The algorithm initializes the distance (*dist*) of all the vertices to infinity and predecessor (*pred*) in the path to *NULL* (Lines 2–5). However, the *dist* value of the source vertex *src* is made zero. While *dist* is useful in finding the shortest distance (in terms of number of edges from the *src* vertex), *pred* is used to find the actual shortest path. The variable *changed* is made `False` at the beginning of the `while()` loop. The outermost *forall* loop iterates over all the vertices in the

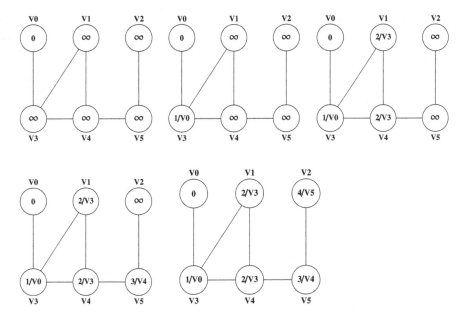

Fig. 2.1 Breadth-first search illustration on an example graph

graph, while the innermost *forall* loop iterates over the neighbours of each vertex (*outnbrs*) of the graph. The BFS distance *dist* of a vertex t is updated to *p.dist + 1* and its predecessor *t.pred* is made p, if the following two conditions are satisfied (see Lines 11–15):

1. There is an edge $p \rightarrow t$
2. $t.dist > p.dist + 1$

Whenever such an update happens, the variable *changed* takes the value True. If no update happens in a particular iteration of the while() loop, the exit condition is satisfied (Line 18) and the algorithm terminates.

Figure 2.1 shows the working of the BFS traversal on an undirected graph with six vertices.[2] Each iteration of the loop is shown as a subfigure. The algorithm halts after five iterations, because there is no update in the fifth iteration. For a graph with n vertices, BFS requires $O(n^2)$ time with an adjacency matrix representation. It may be noted that the BFS algorithm presented above has little resemblance to the BFS algorithms presented in text books, which invariably use a queue of vertices. However, it must be pointed out that Algorithm 2.1 discovers all paths of length k from the vertex *src*, before discovering paths of length $k + 1$, and hence is *breadth-first* in nature. It is more suitable for parallel implementations than the one that uses queues.

---

[2]Note that BFS works well for directed graphs also.

2.2.2 Depth-First Search

The DFS algorithm (Algorithm 2.2) visits each vertex of a graph exactly once starting from a source vertex [29]. The algorithm upon visiting a vertex, selects one of its neighbours and visits that vertex deeper. It also timestamps the vertices, *start* indicating the time at which the vertex was entered (visited) for the first time, and *etime* indicating the time at which the vertex was exited. Colors are used to indicate the state of a vertex: *white* indicating that the vertex has not yet been visited, *gray* during the time (*etime* − *start*), and *black* at *etime* and later. The timestamps are useful to prove certain properties of DFS.

Algorithm 2.2: Depth-First Search Algorithm

```
1  DFS (Graph G) {
2      clock = 0;
3      foreach( Vertex p In G ){
4          p.color = white;
5          p.start = p.etime = p.pred = -1;
6      }
7      foreach( Vertex p In G ){
8          if( p.color == white ){
9              Traverse(p, G);
10         }
11     }
12 }
13 Traverse (Vertex p, Graph G) {
14     p.start = ++clock;
15     p.color = gray;
16     foreach( Vertex t In p.outnbrs ){
17         if( t.color == white ){
18             t.pred = p;
19             Traverse(t, G);
20         }
21     }
22     p.color = black;
23     p.etime = ++clock;
24 }
```

It initializes each vertex with properties *color*=white, *start* and *etime* to −1, and *pred*=−1 (see Lines 3–6). Then each unexplored vertex (i.e *color*=white) is selected and DFS traversal is initiated (see Lines 7–11). In the traversal function *Traverse*() (Lines 13–24), for the (argument) vertex *p*, its *start* time is noted and *color* is made *gray*. This denotes vertex *p* is now visited. Then all the unexplored neighbours (*outnbrs*) of the vertex *p* are visited in a depth-first manner recursively (Lines 16–21) by calling the function *Traverse()* (Line 19). When the recursive call of a vertex returns, the *etime* of the vertex is marked, and the color of the vertex is made *black*. This denotes there is no more vertices to visit using the vertex *p*.

Figure 2.2 shows the DFS traversal on a directed graph.[3] The *start/etime* of the vertices are updated, and the color is updated to gray and then to black. The edge explored in each step is shown as a dashed line and the iterations are shown in Fig. 2.2a–h. DFS induces a spanning forest on a graph. The edges after the DFS traversal of a directed graph can be classified as *tree, forward, back* or *cross* edges:

- An edge $p \to t$ is a *tree edge* if vertex t was unexplored when it was visited by traversing $p \to t$. All tree edges belong to the spanning forest.
- An edge $t \to p$ is a *back edge* if it connects vertex t to an ancestor p in the spanning forest.[4]
- An edge $p \to t$ is a *forward edge* if it is not a tree edge and connects vertex p to a proper descendant t in the spanning forest.
- All other edges are classified as *cross edges*.

The edge $v0 \to v3$ is a *tree edge*, edge $v1 \to v3$ is a back edge and $v2 \to v4$ is a *cross edge* in Fig. 2.2h. DFS on an undirected graph results in only tree and back edges.

The DFS algorithm is used in finding connected and strongly connected components, solving puzzles, finding biconnectivity in graphs, and solving a host of other problems related to graphs. As an example, Algorithm 2.3 shows how DFS can be used to find connected components of an undirected graph. Figure 2.3 shows connected components of a disconnected graph. assuming that DFS on the disconnected graph in Fig. 2.3 begins at vertex 1, the order of visiting the vertices with DFS would be 1, 7, 8, 2 (forming component 0), 3, 9, 4, 10 (forming component 1) and 5, 6, 11, 12 (forming component 2). The connected components found are always independent of the vertex from which DFS is started.

DFS is often used in gaming simulations where each choice or action leads to another, yielding a choice tree. It traverses the choice tree until it discovers an optimal solution path (e.g., win). The choice tree may not be built explicitly. For a graph with n vertices, DFS requires $O(n^2)$ time with an adjacency matrix representation.

2.2.3 Single Source Shortest Path (SSSP)

An algorithm to solve SSSP problem takes a weighted directed graph and a distinguished source vertex *src* as inputs, and computes the cost of the *shortest* path from *src* to all other vertices in the graph [30]. The sum of the edge weights in a path constitutes the path's cost. Algorithm 2.4 shows the pseudo code for SSSP computation. *dist* and *pred* properties are used to store the weight of the shortest

---

[3]Note that DFS works well for undirected graphs also.

[4]Node a is an ancestor of node b in a rooted tree, if there is a directed path from a to b. b is a proper descendant of a in the same context, if $a \neq b$.

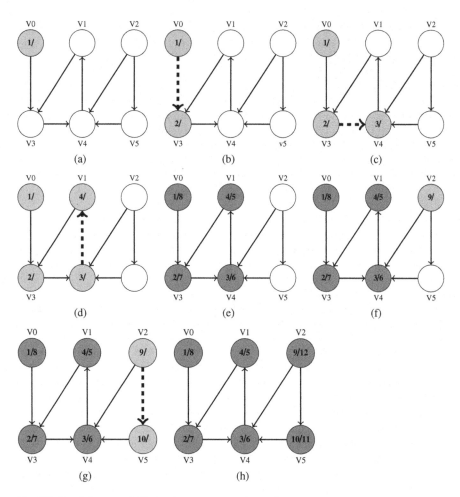

Fig. 2.2 Depth-first search illustration on an example graph

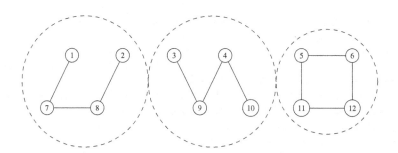

Fig. 2.3 Example of connected components. Each dashed circle contains one connected component. The given graph is the union of the three components

---

Algorithm 2.3: Connected Components Algorithm Using DFS

```
 1  DFS_CC (Graph G) {
 2      clock = 0; cn = 0;
 3      foreach( Vertex p In G ){
 4          p.color = white;
 5          p.component = 0;
 6          p.start = p.etime = p.pred = -1;
 7      }
 8      foreach( Vertex p In G ){
 9          if( p.color == white ){
10              Traverse(p, G, cn);
11              cn = cn + 1;
12          }
13      }
14  }
15  Traverse (Vertex p, Graph G, int Comp_num) {
16      p.start= ++clock;
17      p.component = Comp_num;
18      p.color = gray;
19      foreach( Vertex t In p.outnbrs ){
20          if( t.color == white ){
21              t.pred = p;
22              Traverse(t, G, Comp_num);
23          }
24      }
25      p.color = black;
26      p.etime = ++clock;
27  }
```

---

path and predecessor vertex in the shortest path respectively for each vertex. *dist* and *pred* are initialized to ∞ and *NULL* respectively for all the vertices (see Lines 2–5). However, *dist* value of the source vertex *src* is initialized to zero.

The *dist* value of vertices is reduced by traversing all the edges in the graph multiple times (see Lines 7–20) until a fixpoint is reached. At the beginning of each iteration, the variable *changed* is set to `False`. The variable *changed* is set to `True` whenever *dist* of *any* vertex is reduced during an iteration of the `while` loop (see Line 14). The shortest distance value of a vertex *t* is reduced when it is cheaper to go to vertex *p*, and then traverse the edge $p \rightarrow t$ (if there is one), than using the already computed path to *t*. That is, $t.dist > (p.dist + weight_{edge(p,t)})$ (Lines 11–15). This triggers one more iteration of the `while` loop. The exit condition is satisfied when there is no modification to *dist* value of *any* vertex in an iteration (see Line 19). This marks the end of the algorithm. The shortest path to a vertex *p* can be computed using *pred*, by simply traversing backwards from any vertex *p*.

There are many algorithms for finding the shortest paths. The popular ones are the *Bellman–Ford* algorithm, *Dijkstra's* algorithm and the Δ-stepping algorithm. The Bellman–Ford algorithm can detect negative cycles in a graph and it has a complexity of $O(m.n)$, where *m* and *n* are the number of edges and vertices

---

Algorithm 2.4: Single Source Shortest Path Computation

```
 1  SSSP(Graph G, Vertex src) {
 2      foreach( Vertex p In G ){
 3          p.dist = ∞;
 4          p.pred = NULL;
 5      }
 6      src.dist = 0;
 7      while( True ){
 8          changed = False;
 9          foreach( Vertex p In G ){
10              foreach( Vertex t In p.outnbrs ){
11                  if( t.dist > (p.dist + G.getWeight(p, t)) ){
12                      t.dist = p.dist + G.getWeight(p, t);
13                      t.pred = p;
14                      changed = True;
15                  }
16              }
17          }
18      }
19      if ( changed == False ) break;
20  }
```

---

respectively. Dijkstra's algorithm works for graphs with non-negative edge weights only, and has a complexity of $O(m \ log(n))$ (with a binary heap) where m is the number of edges. The Δ-stepping algorithm is very efficient for graphs with large diameters. Algorithm 2.4 is a variant of the Bellman–Ford algorithm. SSSP computation is often applied to automatically obtain driving directions between locations in Maps. It is used extensively as a part of many other algorithms, such as Betweenness Centrality computation.

Figure 2.4 shows an example of SSSP computation with Algorithm 2.4 on a graph with source vertex S. The *dist* and *pred* values are shown inside the circles corresponding to vertices as (*dist/pred*) in the figure.

2.2.4 All Pairs Shortest Path Computation (APSP)

Apart from SSSP, it may be necessary to compute shortest paths between every pair of distinct vertices in a directed graph. Algorithm 2.5 achieves this and this algorithm is also called *Floyd's Algorithm* or *Floyd-Warshall Algorithm* [31] for computing all pairs shortest paths. The principle of this algorithm can be enunciated using the following equations and Fig. 2.5:

$$A^0[i, j] = C[i, j].$$ This is cost of the edge(i, j)

$$A^k[i, j] = min\{A^{k-1}[i, j], A^{k-1}[i, k] + A^{k-1}[k, j]\}, k \geq 1.$$

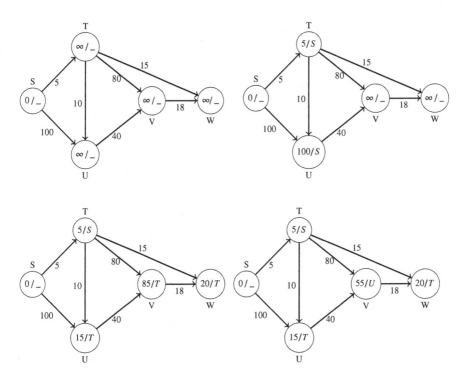

Fig. 2.4 SSSP Computation illustration on an example graph (Text inside a vertex indicates current distance and the current predecessor)

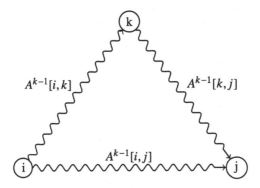

Fig. 2.5 Principle of Floyd's algorithm

Algorithm 2.5: All Pairs Shortest Path Computation

```
1  APSP(Graph G) {
2        //For each vertex v, v.dist[p] stores the shortest distance to vertex p from v, ∀ p, v ∈
         G.V.
3        //Similarly, v.pred[p] stores the highest numbered intermediate vertex through which
         the
4        //shortest path from v to p passes ∀ p,v ∈ G.V.
5        foreach( Vertex v In G ){
6              foreach( Vertex p In v.outnbrs ){
7                    v.dist[p] = G.getWeight(v,p);
8                    v.pred[p] = NULL;
9              }
10             v.dist[v]=0; v.pred[v]=v;
11       }
12       foreach( Vertex k In G ){
13             foreach( Vertex i In G ){
14                   foreach(  Vertex j In G ){
15                         if ( i.dist[k]+k.dist[j] <i.dist[j] ){
16                               i.dist[j] = i.dist[k]+k.dist[j], i.pred[j] = k;
17                         }
18                   }
19             }
20       }
21 }
```

Where, $A^k[i, j]$ is the cost of the shortest path from vertex i to vertex j without going through any vertex numbered higher than k. In going from vertex i to vertex k and then from vertex k to vertex j, we do not go through vertex k or any vertex numbered higher than k.[5] The cost here is as in SSSP computation (sum of edge costs). Computation of matrix A can be performed as shown in Algorithm 2.5, where each row of A is attached to the corresponding vertex of G as a property. Floyd's algorithm has a time complexity of $O(n^3)$ for a graph with n nodes. Figures 2.6 and 2.7 show an input graph and the distance matrices. Examples of computing the entries in the matrices are provided below.

$$A^1[v_3, v_2] = \min\{A^0[v_3, v_2], A^0[v_3, v_1] + A^0[v_1, v_2]\} = \min\{8, 5 + 2\} = 7$$

$$A^2[v_1, v_3] = \min\{A^1[v_1, v_3], A^1[v_1, v_2] + A^1[v_2, v_3]\} = \min\{\infty, 2 + 3\} = 5$$

$$A^3[v_2, v_1] = \min\{A^2[v_2, v_1], A^2[v_2, v_3] + A^1[v_3, v_1]\} = \min\{\infty, 3 + 5\} = 8$$

---

[5]Reaching vertex k once and leaving vertex k once is obviously permitted here.

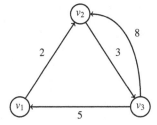

Fig. 2.6 Example of APSP computation: input graph

A^0	v_1	v_2	v_3		A^1	v_1	v_2	v_3		A^2	v_1	v_2	v_3		A^3	v_1	v_2	v_3
v_1	0	2	∞		v_1	0	2	∞		v_1	0	2	5		v_1	0	2	5
v_2	∞	0	3		v_2	∞	0	3		v_2	∞	0	3		v_2	8	0	3
v_3	5	8	0		v_3	5	7	0		v_3	5	7	0		v_3	5	7	0

Fig. 2.7 Example of APSP computation: distance matrices

2.2.5 Strongly Connected Components (SCC)

Strongly connected components can be found by running multiple depth-first traversals: one on the original graph and another on its transpose [32]. The pseudo code for computing SCC is shown in Algorithm 2.6. The *transpose* of a directed graph $G(V, E)$ is denoted by $G^T(V, E^T)$ where $E^T = \{(v, u) : (u, v) \in E\}$. The algorithm takes a directed graph G(V,E) as input, and uses Depth First Search (DFS) to compute the end time (*etime*) for each vertex v $\in V$. This is Step 1. G^T is computed in Step 2. In Step 3, the vertices of the graph are sorted in the decreasing order of *etime*. A DFS on G^T is initiated in the sorted order of the vertices (Step 4). This is the order now prescribed for visiting vertices in the forall loop (see Lines 7–11 in Algorithm 2.2). The order in which the neighbours of a selected vertex are visited is immaterial (see Lines 16–21 in Algorithm 2.2). The vertices of each spanning tree generated during DFS traversal on G^T correspond to vertices of an SCC of the input graph. Each SCC may then be created by taking the vertices of the corresponding spanning tree and the edges between these vertices from the input graph. The SCCs are immune to the order in which DFS is invoked on the vertices in Step 1. The time complexity of this SCC algorithm is the same as that of DFS.

Figure 2.8 shows the execution of Algorithm 2.6 on an example graph. The figure is self-explanatory. Algorithm 2.6 is simple to understand but in practice, it is not very efficient due to two DFS traversals involved. There is an optimal algorithm due to Tarjan [33] which involves only one DFS. There is a third algorithm that uses the divide and conquer strategy and is shown in Algorithm 2.7.[6] This algorithm computes vertices which are reachable in forward and backward directions using

---

[6]This algorithm exhibits more parallelism than that exhibited by Tarjan's algorithm.

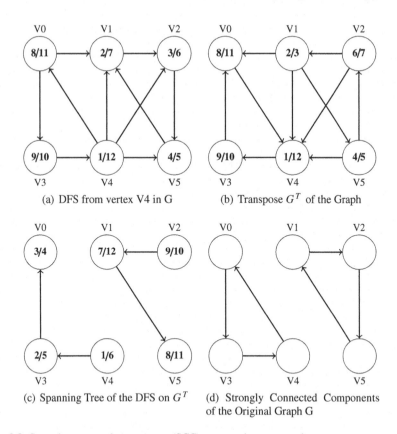

(a) DFS from vertex V4 in G (b) Transpose G^T of the Graph

(c) Spanning Tree of the DFS on G^T (d) Strongly Connected Components of the Original Graph G

Fig. 2.8 Strongly connected components (SCC) computation on a graph

---

Algorithm 2.6: Strongly Connected Components Algorithm

---

1 *SCC(Graph G)* {
2   Step 1: Call DFS(G) and compute *v.etime* for each vertex $v \in V$.
3   Step 2: Compute G^T.
4   Step 3: Sort the vertices in the decreasing order of the value of *etime*.
5   Step 4: Call $DFS(G^T)$ using sorted order of vertices.
6   Step 5: Each spanning tree resulting from the depth-first search in Step 4 corresponds to an SCC.
7 }

BFS traversals on the graph G and its transpose, G^T. The algorithm is based on the following lemma from [34]:

Let G = (V, E) be a directed graph and let $i \in V$ be a vertex in G. Then $V_{fwd}(i) \cap V_{bwd}(i)$ is a unique SCC in G. Moreover, for every other SCC s in G, exactly one of the following holds:

- $s \subset (V_{fwd}(i) \setminus V_{bwd}(i))$.
- $s \subset (V_{bwd}(i) \setminus V_{fwd}(i))$.
- $s \subset (V \setminus (V_{fwd}(i) \cup V_{bwd}(i)))$.

The lemma states that for any vertex $i \in G$, its SCC can be identified with the intersection set of the two sets, $V_{fwd}(i) \cap V_{bwd}(i)$. Here, $V_{fwd}(i)$ is the set of vertices reachable from i (forward reachable set) and $V_{bwd}(i)$ is the set of vertices that can reach i (backward reachable set). Further, it also asserts that any other SCC s of G falls entirely within purely forward reachable set $(V_{fwd}(i) \setminus V_{bwd}(i))$, or purely backward reachable set $(V_{bwd}(i) \setminus V_{fwd}(i))$, or purely non-reachable set $(V \setminus (V_{fwd}(i) \cup V_{bwd}(i)))$ of vertices. The forward and the backward reachable sets can be found with BFS on the graph G and its transpose. Note that transpose of G indicates incoming edges of each vertex. This algorithm achieves an expected time complexity of $O(e \log n)$.

As an example for Algorithm 2.7, consider the graph G in Fig. 2.8a, and the graph in Fig. 2.8b, which shows G^T, the transpose of G (ignore the numbers inside the nodes). The forward reachable set from vertex $V_0 = \{V_0, V_1, V_2, V_3, V_4, V_5\}$ (computed using BFS on G) and backward reachable set from vertex $V_0 = \{V_0, V_4, V_3\}$ (computed using BFS on G^T). Intersection of these sets forms an SCC which is the subgraph with vertices $\{V_0, V_4, V_3\}$ and edges between these vertices. These vertices are removed from the graph G, which results in graph having only vertices $V = \{V_1, V_2, V_5\}$ and edges which connect these vertices. Forward and backward reachable sets on the new graph and its transpose are the sets $\{V_1, V_2, V_5\}$ and $\{V_1, V_2, V_5\}$ respectively. The intersection of these sets forms an SCC

Algorithm 2.7: Divide and Conquer Algorithm for Finding Strongly Connected Components

---

```
1  SCC(Graph G, Vertexset P, Vertexsetcollection StrongCompSet) {
2      // P is the vertex set on which forward and backward searches are performed.
3      // StrongCompSet is the collection of strongly connected components,
4      // represented by their respective vertex sets.
5      if ( P is empty ){
6          return;
7          //End of recursion.
8      }
9      Select v uniformly at random from P;
10     // v is the pivot vertex from which forward and backward searches are made.
11     V_fwd = Fwd-Reachable(G, P, v); // This is a BFS on G, starting from v.
12     V_bwd = Fwd-Reachable(G^T, P, v); // This is a BFS on G^T, starting from v.
13     scc = V_fwd ∩ V_bwd;
14     StrongCompSet = StrongCompSet ∪ scc;
15     SCC(G, V_fwd \ scc, StrongCompSet); // Recursive call.
16     SCC(G, V_bwd \ scc, StrongCompSet); // Recursive call.
17     SCC(G, V_G \ (V_fwd ∪ V_bwd), StrongCompSet); // Recursive call. V_G is the vertex set
           of G.
18  }
```

which is the subgraph with vertices $\{V_1, V_2, V_5\}$ and edges between these vertices. The two SCCs are shown in Fig. 2.8d.

The SCC algorithm is used as a preprocessing step in many algorithms to form clusters of vertices. It has applications in model checking, vector code generation in compilers, and analysis of transportation networks.

2.2.6 Weakly Connected Components (WCC)

A *directed* graph G is *weakly connected* if the undirected graph G', obtained after discarding the edge directions in G, is connected. WCC of G are the connected components of G' and can be found by running the DFS (Algorithm 2.2) on G'. It may be noted that while WCC of the graph in Fig. 2.8 is the original graph without the directions of the edges, the SCC of the same graph are different. This algorithm is used to find disconnected clusters in graphs, usually in a preprocessing step of other algorithms.

2.2.7 Minimum Spanning Tree (MST)

In this section, we deal with only connected undirected graphs. A graph can have more than one spanning tree. The *cost* of a spanning tree is the sum of the weights of all the edges in the spanning tree. A spanning tree which has the minimum cost is called a *Minimum Spanning Tree (MST)*. A weighted graph can have more than one MST with the same weight but with different sets of edges. MSTs are not necessarily rooted trees, i.e., they may not have a vertex identified as a root.

There are several algorithms to compute MST, such as Prim's [35], Kruskal's [36] and Boruvka's [37]. Prim's algorithm follows a greedy strategy. It starts from an arbitrary vertex, considered as a component, and adds the minimum weight edge incident on the component from the non-component vertices. The component thus grows, and this step of choosing the minimum weight edge from the current component is repeated. When the component contains all the vertices, the set of edges considered during the processing forms the MST.

Kruskal's algorithm to find an MST of a graph G is as follows: Sort the edges of G by their edge weights. Add the edges to the MST[7] (initially empty) in the increasing order of edge weights, such that adding an edge does not lead to a cycle. Stop once $|V| - 1$ edges are added to the tree or all the edges are considered. Kruskal's algorithm uses a *Union-Find* data structure for storing the intermediate forest.

---

[7]In intermediate stages of the algorithm, we may have a forest and not a tree. A tree is obtained usually only at the end.

Algorithm 2.8: Boruvka's Minimum Spanning Tree Algorithm

```
1  MST(Graph G) {
2      Union_Find set = { G.V};
3      foreach( t In set ){
4          |   t.parent = t;
5      }
6      MST = φ;
7      mst_cost = 0;
8      while( set contains more than one component ){
9          foreach( Vertex p In G ){
10             Step 1: Find the edge e : p ↔ t with minimum weight and p.parent ≠
                   t.parent
11             // e connects subtree corresponding to p and
12             // subtree corresponding to t
13             Step 2: add e to MST
14             Step 3: Union the components p and t in set using Union(p, t)
15         }
16     }
17     Step 4: mst_cost = sum of edge weights of all e ∈ MST
18     Step 5: output MST and value mst_cost
19 }
```

Algorithm 2.8 shows Boruvka's MST algorithm [37, 38] which also uses a *Union-Find* data structure for storing disjoint sets. Each set is identified with an identifier, called *parent* (say). The operations on the data structure are union(A, B) and find(x). The union(A,B) operation makes a destructive[8] union of sets A and B into a single set, and sets the *parent* of both A and B to the same value. It is important to note that the parent of a set is a representative of the set and is not necessarily the parent of all the vertices of the tree which the set represents. The find(x) operation returns the *parent* of the set to which the element x belongs. The *set* variable is initialized as a *Union-Find* data structure where each point is a separate set (see Line 4). The set of edges in the *MST* is initialized to the *null* set (see Line 6) and *mst_cost* is initialized to zero (see Line 7).

The while loop (Lines 8–16) computes the *MST*. The edge $e : p \leftrightarrow t$ is added to the MST if e connects two disconnected subtrees (p.parent ≠ t.parent) and the weight of e is the minimum of the edges that connect the two subtrees (Line 10). This edge is added to the MST (Line 13) and the two subsets are unified to form a single subset (Line 14). The while loop terminates when *set* contains exactly one component (assuming that the input graph is connected).

Boruvka's algorithm has a time complexity of $O(m \log n)$, where n and m are the number of vertices and number of edges (respectively) in the graph. Boruvka's algorithm is easier to parallelize than either Prim's or Kruskal's algorithms. MST computation is an important and basic problem with diverse applications which

---

[8]Originals are destroyed.

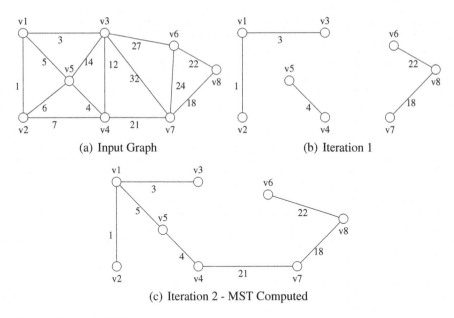

(a) Input Graph   (b) Iteration 1

(c) Iteration 2 - MST Computed

Fig. 2.9 Boruvka's MST computation

include network design, approximation algorithms for hard problems, cluster analysis, etc.

Figure 2.9 shows the working of Boruvka's algorithm on a sample graph. The input graph is shown in Figure 2.9a. It may be noted that vertices $v1$ and $v2$ choose the same minimum weight edge ($v1 \leftrightarrow v2$). After the first iteration, vertices $\{v1, v2, v3\}$ form a subtree with the same parent ($v3$). Similarly the subsets of vertices $\{v4, v5\}$ and $\{v6, v7, v8\}$ are merged and the vertices $v5$ and $v8$ (respectively) are made their parents (see Fig. 2.9b for the partial MST with three trees). In the second iteration, the MST is computed by adding two minimum weight edges $v1 \leftrightarrow v5$ and $v4 \leftrightarrow v7$, which connect the three disjoint subsets. The edges that connect vertices in the same tree are not chosen. After this iteration, all the vertices merge into a single set with the *parent* being vertex $v8$. Figure 2.9c shows the MST.

2.2.8   Maximal Independent Set (MIS) Computation

In an undirected graph $G(V, E)$, a set $I \subseteq V$ is called *independent*, if for every pair of vertices $u, v \in I$, $(u, v) \notin E$. That is no two vertices in I are adjacent in G. An independent set I is *maximal*, if $\forall p \in (V - I)$, $p \cup I$ is not independent. The independent set I is *maximum* for the graph $G(V, E)$, if I is an independent set of the largest possible size. Algorithm 2.9 shows how a *maximal* independent set is computed. Whenever a vertex v is added to the independent set I, all the *sink* vertices of edges with v as the source vertex are removed from V_m. MIS is one of the important problems in the study of graph algorithms, because several other fundamental graph problems can be reduced to MIS with a constant round complexity overhead. Maximal matching, vertex and edge coloring, vertex cover, and approximation of maximum matching are some of them.

Finding a maximum independent set (MXMIS) for a graph is a computationally hard problem. Good approximation algorithms exist for finding MXMIS in *planar graphs*. The MXMIS problem is closely related to cliques and vertex cover in graphs. MXMIS computation is useful in genetic algorithms, neural networks, and location theory.

Algorithm 2.9: Maximal Independent Set Algorithm

```
1  Maximal_Independent_Set(Graph G) {
2      I = φ, Vm = V;
3      while( Vm ≠ φ ){
4          choose v ∈ Vm;
5          I = I ∪ v;
6          Vm = Vm − (v ∪ Neighbours(v));
7      }
8      return I;
9  }
```

2.3 Other Algorithms

This section describes some of the important problems that use graphs and graph related algorithms. The list is only a sampler and is not exhaustive. The description is intended to provide only an introduction to the problems.

2.3.1 Pagerank Algorithm

The Pagerank algorithm [39, 40] is used to index webpages, and decide the quality of a webpage. Webpages are given values based on the number of incoming links to a page and weight of each linking page (i.e., *source* of the link). It is also used for ranking text for entity relevance in natural language processing.

The page rank of a page or website A is calculated by the formula given below:

$$PR(A) = (1 - d) + d \times (PR(T_1) \div C(T_1) + \ldots + PR(T_n) \div C(T_n))$$

Algorithm 2.10: PageRank Algorithm Pseudo Code

```
 1  d = 0.85; // damping factor
 2  pagerank(Vertex P, Graph G) {
 3      val = 0.0;
 4      foreach( Vertex t In p.innbrs ){
 5          if( t.outdegree != 0 ){
 6              | val += t.pr ÷ t.outdegree;
 7          }
 8      }
 9      p.pr = val × d + (1 - d) ÷ |G.V|;
10  }
11  computePR(Graph G, MAX_ITR) {
12      foreach( t In G.V) t.pr = 1 ÷ |G.V|;
13      itr = 1;
14      while( itr < MAX_ITRS ){
15          foreach( Vertex p In G ){
16              | pagerank(p, G);
17          }
18          itr++;
19      }
20  }
```

T_1 to T_n correspond to the websites accessed by the user. PR(A) is the page rank of A. PR(T_1) to PR(T_n) are the page ranks of the pages which have *urls* linking to A. C(T_1) to C(T_n) are the number of links on the sites T_1 to T_n respectively. The parameter d in the computation is called the damping factor and is typically set to 0.85. The damping factor is the probability at each page, that a "random surfer" will get bored and request another random page.

Algorithm 2.10 shows the pseudo-code to compute the page ranks of webpages. The vertices of the graph G in the pagerank algorithm are the websites accessed by the user. The incoming *urls* from a webpage A to webpage B is represented as the edge $A \rightarrow B$. Similarly, the edge $B \rightarrow A$ represents the outgoing *urls* to a website A from B. The page rank of a vertex is high when it has more incoming edges. The function *pagerank()* (Lines 2–10) updates the page rank of a vertex using the number of incoming edges to it and the outdegree of the *source* vertex of the incoming edge. The function *computePR()* (Lines 11–20) calls the function *pagerank()* (Line 16 MAX_ITR times. This *heuristic* indexes webpages efficiently.

2.3.2 Triangle Counting

Triangle is an elementary subgraph used in the analysis of complex networks and graphs. A simple algorithm to find the number of triangles in a graph is shown in Algorithm 2.11. It has applications in social network analysis as well as geometry.

Algorithm 2.11: Triangle Counting Algorithm

```
 1  TC(Graph G) {
 2      tc = 0;
 3      foreach( Vertex p In G ){
 4          foreach( t In p.nbrs ){
 5              foreach( r In p.nbrs ){
 6                  if( t≠ r && t ∈ r.nbrs ){
 7                      ++tc;
 8                  }
 9              }
10          }
11      }
12      G.triangles = tc / 6;
13  }
```

2.3.3 Graph Coloring

Graph coloring [41] is the assignment of colors to the vertices or edges of a graph. A coloring of a graph such that no two adjacent vertices share the same color is called a vertex coloring of the graph. Similarly, an edge coloring assigns a color to each edge so that no two adjacent edges share the same color. A coloring using at most k colors is called k-coloring. Graph coloring has applications in process scheduling, register allocation phase of a compiler, and also in pattern matching. Theoretically, it is an *NP-Complete* problem for general graphs, and many heuristics have been proposed in literature for this problem. Greedy vertex coloring is one such heuristic and it is described in Algorithm 2.12.

The greedy algorithm [42] considers the vertices of the graph in a specific order that is already chosen, say, $v_0, v_1, \ldots, v_{n-1}$. Colors are numbered as $1, 2, \ldots$. Vertex v_i is assigned the lowest numbered available color that is not used by its predecessors among $v_0, v_1, \ldots, v_{i-1}$, which are also neighbors of v_i. The quality of the computed coloring depends on the chosen ordering of vertices. While there exists an ordering of vertices which when used with the just described greedy coloring leads to an optimal number of colors, the number of colors can be much larger than the optimal for arbitrary orderings. One popular ordering of vertices is the ordering by their degree: *largest degree first*. Another is to choose a vertex v of minimum degree, find the ordering for the subgraph with v removed recursively, and then place v last in the ordering. *Degeneracy* of the graph is the largest degree d encountered during the execution of this algorithm as the degree of a removed vertex. This order is called *smallest degree last* and coloring with this ordering uses at most $d + 1$ colors. Figure 2.10 shows an example of coloring using the greedy strategy with different orders.

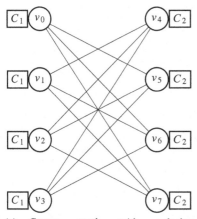

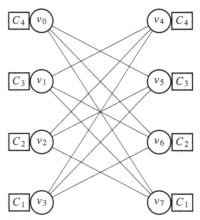

(a) Crown graph with coloring generated by the vertex order: $v_0, v_1, v_2, v_3, v_4, v_5, v_6, v_7$. Two colors C_1 and C_2 are required (shown in brackets beside vertices).

(b) Crown graph with coloring generated by the *largest degree first* vertex order: $v_3, v_7, v_2, v_6, v_1, v_5, v_0, v_4$. Four colors C_1, C_2, C_3 and C_4 are required (shown in brackets beside vertices).

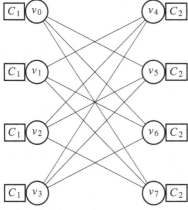

(c) Crown graph with coloring generated by the *smallest degree last* vertex order: $v_4, v_2, v_3, v_5, v_6, v_1, v_7, v_0$. Two colors C_1 and C_2 are required (shown in brackets beside vertices).

Fig. 2.10 Example of graph coloring

2.3.4   K-Core

A K-Core [43] of a graph G is a maximal connected subgraph H of G, with all the vertices in H having minimum degree K. The basic algorithm is to first remove all the vertices $v \in G$ with degree <K. This will reduce the degree of the vertices which are adjacent to v. The process is repeated until a fixpoint is reached. The remaining

---

Algorithm 2.12: Greedy Graph Coloring Algorithm

```
 1  GColor(Graph G, Sequence Order) {
 2  |    // Greedy vertex coloring for undirected graphs.
 3  |    // For each vertex in G, v.color is its assigned color.
 4  |    // Colors are numbered 1, 2, 3, ....
 5  |    // Order is the sequence of vertices and establishes the
 6  |    // order in which vertices will be considered for coloring.
 7  |    // For example, Order = v₀, v₁, v₂, ..., vₙ₋₁.
 8  |    // The sequence, 0, 1, 2, ..., (n − 1) is used to index into Order.
 9  |    // Order.i is the iᵗʰ vertex in the sequence, vᵢ and Order.i.color is its color.
10  |    foreach( Vertex v In G ){
11  |    |    v.color = 0;// Initialization of color
12  |    |    foreach( i In 0, 1, ..., (n − 1) ){
13  |    |    |    AssignColor(i); // Color vertex Order.i;
14  |    |    }
15  |    }
16  }
17  AssignColor(int k) {
18  |    Vertex v = Order.k; v.color = 1; // Try color = 1 first.
19  |    // Now check all k-1 vertices in Order to see
20  |    // if color 1 is already taken; if so, try the next color.
21  |    // If all colors 1, 2, ..., k are taken, v is assigned color k+1
22  |    foreach( i In 0, 1, ..., (k − 1) ){
23  |    |    if ( (Order.i In v.nbrs)&&(v.color == Order.i.color) ){
24  |    |    |    v.color = Order.i.color + 1 ;
25  |    |    }
26  |    }
27  }
```

---

graph is either K-core or empty. The algorithm has applications in areas such as study of the clustering structure of social network graphs, network analysis, and computational biology.

2.3.5 *Betweenness Centrality (BC)*

For every pair of vertices (u, v) of an undirected weighted graph, there could be many shortest paths between u and v. For each vertex x, the *Vertex Betweenness Centrality* (BC) [44] is the number of shortest paths that pass through x.[9] With this definition, it is unbounded. To make it bounded, it is normalized as:

$$BC(v) = \sum_{s \neq v \neq t} \frac{SP_{st}(v)}{SP_{st}} \tag{2.1}$$

---

[9]Edge Betweenness Centrality is defined towards the end of this section.

where, $BC(v)$ is the Vertex Betweenness Centrality of vertex v, SP_{st} is the number of shortest paths (*without considering the weights*[10]) between vertices s and t, $SP_{st}(v)$ is the number of these shortest paths passing through v. BC is used in several problems in network theory, including those in social networks, biology, transport and scientific cooperation. Simple algorithms to compute BC use a modified form of the *Floyd-Warshall* algorithm to find all pairs shortest paths, and thereby require $\Theta(n^3)$ time and $\Theta(n^2)$ space, for an undirected graph with n vertices. While this is acceptable for dense graphs, for real life graphs which are relatively sparse, a very well known algorithm is due to Brandes [45]. This requires $O(n.e)$ time and $O(n + e)$ space, where n and e are the number of vertices and edges (respectively) in the unweighted undirected graph.

Algorithm 2.13 shows the modifications necessary to the *Floyd-Warshall* to compute vertex betweenness centrality. Updating SP_{ij} can be done while updating the shortest distance between the vertices i and j, i.e. $i.dist[j]$. While processing shortest paths through vertex k, if a new shortest path is found, then the old SP_{ij} is discarded and the new one is computed as $SP_{ik} * SP_{kj}$, since all shortest paths from vertex i to vertex k can be paired with a shortest path from vertex k to vertex j (see Line 16). Others,wise, if the shortest path through vertex k is of the same length as SP_{ij}, then the new shortest paths through vertex k, ($SP_{ik} * SP_{kj}$) are added to SP_{ij} (see Line 19).

Let $d(s, t)$ be the shortest distance between vertex s and vertex t. $SP_{st}(v)$ is the number of distinct paths with the distance $d(s, t)$. If v is in some shortest path SP_{st}, then $d(s, t) = d(s, v) + d(v, t)$. With this important observation, and having computed $d(s, t)$ and SP_{st} for all vertex pairs (s, t), $SP_{st}(v)$ can now be computed as (see Line 29):

$$SP_{st}(v) = SP_{sv} * SP_{vt}, \texttt{ if } d(s, t) == d(s, v) + d(v, t) \texttt{ else } 0. \qquad (2.2)$$

Vertex Betweenness Centrality is computed in Line 24 using Eq. 2.1.

Vertices with high BC may have considerable control over information passing between others, and their removal from the network will disrupt communications between other vertices to the maximum extent because they lie on the largest number of paths taken by messages. Two examples of vertex BC computation are shown in Fig. 2.11. Consider vertex 5 in graph (1) in Fig. 2.11a. The shortest paths (SP) between pairs of vertices which include vertex 5 are required to be computed. SP between 1 and 2, 1 and 3, 1 and 4, and 1 and 6 do not include vertex 5. Only the SP between 1 and 7 includes vertex 5. Vertices 2 and 3 also contribute one path each to BC. Vertices 4 and 6 contribute two paths each to BC, and vertex 7 contributes five paths to BC. Thus BC(5) = 12. That means vertex 5 is very important to this network. Graph (2) in Fig. 2.11c contains multiple shortest paths between several pairs of vertices. For example, there are two shortest paths from vertex 2 to vertex 3, one passing through vertex 1 and another passing through vertex 4. Therefore,

---

[10]These are called Geodesic paths.

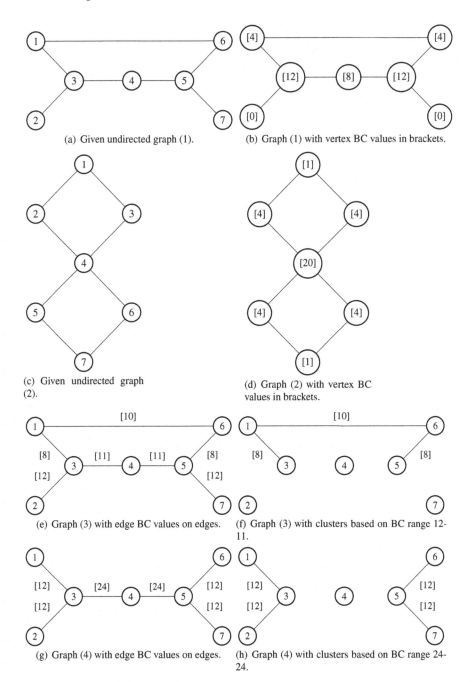

(a) Given undirected graph (1).

(b) Graph (1) with vertex BC values in brackets.

(c) Given undirected graph (2).

(d) Graph (2) with vertex BC values in brackets.

(e) Graph (3) with edge BC values on edges.

(f) Graph (3) with clusters based on BC range 12-11.

(g) Graph (4) with edge BC values on edges.

(h) Graph (4) with clusters based on BC range 24-24.

Fig. 2.11 Examples of betweenness centrality computation

Algorithm 2.13: Vertex Betweenness Centrality Computation

```
1  VertexBetweennessCentrality(Graph G) {
2      //For each vertex v: v.dist[p] stores the shortest distance to vertex p in G,
3      //v.spnum[p] stores the number of shortest paths between vertices v and p, and
4      //v.BC stores the betweenness centrality of vertex v.
5      foreach( Vertex v In G ){
6          foreach( Vertex p In v.outnbrs ){
7              v.dist[p] = 1;
8              v.spnum[p] = 1;
9          }
10     }
11     foreach( Vertex k In G ){
12         foreach( Vertex i In G ){
13             foreach( Vertex j In G ){
14                 if ( i.dist[k]+k.dist[j] <i.dist[j] ){
15                     i.dist[j] = i.dist[k]+k.dist[j];
16                     i.spnum[j] = i.spnum[k]*k.spnum[j];
17                 }
18                 }else if { i.dist[k]+k.dist[j] == i.dist[j] ){
19                     i.spnum[j] = i.spnum[j] + (i.spnum[k]*k.spnum[j]);
20                 }
21             }
22         }
23     }
24     foreach( Vertex v In G ){
25         v.BC = 0;
26         foreach( Vertex s In G ){
27             foreach( Vertex t In G ){
28                 if ( s ≠ t ≠ v ){
29                     if ( s.dist[t] == s.dist[v] + v.dist[t] ){
30                         spnum_v = s.spnum[v] * v.spnum[t];
31                     }
32                     else{
33                         spnum_v = 0;
34                     }
35                 }
36                 delta_BC = spnum_v / s.spnum[t];
37                 v.BC = v.BC + delta_BC;
38             }
39         }
40     }
41  }
```

while computing BC of vertex 4, a contribution of (1/2) will be counted from these two paths.

BC can be similarly defined for edges also, as the number of shortest paths that contain the edge. More formally, it is defined as:

$$BC(e) = \sum_{s \neq t} \frac{SP_{st}(e)}{SP_{st}} \tag{2.3}$$

where, $BC(e)$ is the Betweenness Centrality of edge e, SP_{st} is the number of shortest paths (*without considering the weights*) between vertices s and t, $SP_{st}(e)$ is the number of these shortest paths passing through e.[11] Only minor modifications are needed to Algorithm 2.13 to compute $BC(e)$. It must be observed that for a shortest path from s to t to contain the edge $e(i, j)$, the path must be shortest from s to i and shortest from j to t, and the shortest distance $d(s, t) = d(s, i) + 1 + d(j, t)$. With this observation, and having computed $d(s, t)$ and SP_{st} for all vertex pairs (s, t), $SP_{st}(e(i, j))$ can now be computed as :

$$SP_{st}(e(i, j)) = \begin{cases} SP_{si} & \text{if } (t == j \wedge i \neq j). \\ SP_{jt} & \text{if } (s == i \wedge i \neq j). \\ 1 & \text{if } (s == i \wedge t == j \wedge i \neq j). \\ SP_{si} * SP_{jt} & \text{if } (s \neq t \wedge s \neq i \wedge i \neq j \wedge j \neq t) \\ & \wedge (d(s, t) == d(s, i) + 1 + d(j, t)). \\ 0 & otherwise. \end{cases}$$

Figure 2.11e shows a graph with edge betweenness centrality values. As an example, consider edge (2,3). The SP between 1 and 2, and the SPs between 2 and 3, 2 and 1, 2 and 4, 2 and 5, 2 and 6, and 2 and 7 include the edge (2,3). Similarly, among the SPs from 3, 4, 5, 6, and 7 to other vertices, there is a contribution of 1 SP from each vertex (containing edge (2,3)), making BC((2,3)) as 12. Figure 2.11g shows another example.

Edge BC can be used for community detection by computing clusters based on it. A range of edge BC values, starting with the highest BC value is used in clustering. Clustering begins with the whole graph, and removing edges in the descending order of BC values yields clusters. For example, in Fig. 2.11e, if the range of BC values is 12–11, the edges with BC values 12 and 11 are removed, yielding the clusters in Fig. 2.11f. Figure 2.11h shows another example of clusters obtained when edges with BC value 24 are removed from the graph in Fig. 2.11g. Vertex BC can also be used for clustering, but it requires duplication of nodes that are removed.

---

[11] Sometimes, BC(e) is divided by two because the algorithms traverse each shortest path twice, once from each end point.

2.4 Applications of Graph Analytics in Different Domains

We now briefly discuss a few applications of graph analytics from a variety of domains such as graph mining, graph databases and natural language processing.[12]

2.4.1 Graph Mining

Graph mining [46] finds application in areas such as Bioinformatics, social network analysis, and simulation studies. Some of the problems of interest are the detection of abnormal subgraphs, edges, or vertices in a graph object. Extraction of communities from graphs, and detecting patterns in a graph such as power-law distribution subgraphs and small-diameter subgraphs, are problems of interest in data mining. The following section describes the application of community detection in bioinformatics.

2.4.1.1 Graph Mining in Bioinformatics

Graphs are often used in bioinformatics [47] for describing the processes in a cell. In such graphs, vertices are genes or proteins, and edges may describe protein-protein interaction, or gene regulation. The idea is to find sets of genes that have a biological meaning. One possibility is to compute graph-theoretically relevant sets of vertices and then determine if they are also biologically meaningful. While connected components provide a simple means of achieving this, they have been found to be not so effective. A more advanced idea is to perform graph clustering: find subgraphs that have a high edge density. Figure 2.12 shows an example of graph clustering.

An edge between different clusters could be on several shortest paths from one cluster to another, compared to an edge inside a cluster, because there are more alternative paths inside a cluster. Betweenness centrality of an edge is a good measure of this feature. Removing edges with the highest betweenness centrality from the graph yields good clusters.

2.4.2 Graph Databases (NoSQL)

NoSQL databases may incorporate several types of data, including graphs. Graph databases, which are a part of NoSQL databases, use graph structures ~~with~~ enhanced

---

[12]Applications of graph analytics in computational geometry will be described in Sect. 6.4.

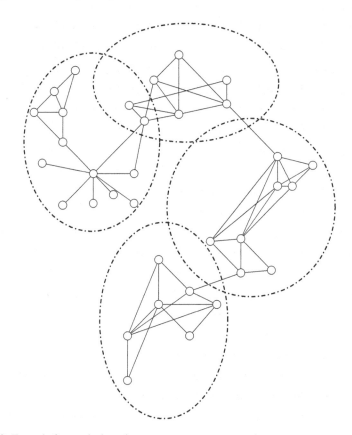

Fig. 2.12 Example for graph clustering

with properties to represent and store data [48]. Answering queries on a graph database requires execution of several graph algorithms on the graph data in the database. Some of the well known graph algorithms, such as SSSP, BFS, DFS, SCC, Betweenness Centrality, Page rank, etc., may be already provided in the query language. An example of a graph database is *Neo4j* with the query language *Cypher* [49].

2.4.3 Text Graphs in Natural Language Processing

Text graphs are used in Natural Language processing and text mining. In text graphs, the textual information is represented using graphs. The vertices may represent a concept in the text, such as paragraphs, sentences, phrases, words, or syllables. An edge between two vertices (concepts) may represent co-occurrence of the two concepts in a window over the text, syntactic relationship, or semantic relationship.

Keyword extraction is an often used application of text graphs. It is used in several day-to-day scenarios, such as in searching using a search engine, indexing and retrieval of books and articles, reviewer assignment for research papers, etc. Common applications are text condensation (paraphrasing a document), information filtering (selection of specific documents of interest), answering keyword based queries, etc. Rousseau et al. [50, 51] and Mihalcea [52] discuss applications of text graphs and their implementation.

Chapter 3
Efficient Parallel Implementation of Graph Algorithms

This chapter provides some insights into the issues in programming parallel algorithms. Parallelism, atomicity, push and pull types of computation, algorithms driven by topology or data, vertex based, edge based and worklist based computation, are some of the important considerations in designing parallel implementations. Parallel algorithms for different types of problems including graph search, connected components, union-find algorithms, and betweenness centrality computation are considered in detail. Finally, a description of the functioning of graph analytics on distributed systems is provided.

3.1 Introduction

Important graph algorithms and their sequential execution were discussed in the previous chapter. This chapter discusses the challenges involved in implementing graph algorithms on heterogeneous distributed systems. Graph processing can be carried out on a multi-core CPU, or on accelerators (e.g., GPUs) connected to a host CPU, or on a distributed system with CPUs and GPUs. Accelerators have a memory that is typically separate from the host CPU memory physically and both have their own address spaces. Data needs to be transferred from the host CPU to the accelerator and vice-versa. The CPU and the GPU follow different execution models with the CPU following a multiple instruction multiple data (MIMD) model, whereas GPUs follow a single instruction multiple thread (SIMT) execution model. A CPU has fewer computing cores and large volatile memory in comparison with a GPU. Table 3.1 shows the major differences between CPU and GPU devices, for a typical configuration. As seen in the table, a CPU has a few tens of powerful (high-frequency) cores. In contrast, a GPU has a few thousand low-frequency cores. Volatile memory of a CPU can be a few hundred GB or even a few TB. In contrast, current GPUs have a limited memory of a few tens of GB. Massive graph analytics

© Springer Nature Switzerland AG 2020
U. Cheramangalath et al., *Distributed Graph Analytics*,
https://doi.org/10.1007/978-3-030-41886-1_3

Table 3.1 CPU and GPU architecture comparison

Device	Make	Cores	Memory	Frequency	Architecture
CPU	Intel Xeon	12	128 GB	2500 MHz	MIMD
GPU	Tesla-K40	2880	12 GB	745 MHz	SIMT

requires distributed computer systems (computer clusters) as the graph is too big to fit in a single device. Each node of such a cluster may have a multi-core CPU and several GPUs. Most clusters are located in a single large box and its nodes are connected with a very high speed interconnection network.

3.2   Issues in Programming Parallel Graph Algorithms

In this section, various issues related to parallel implementations of graph algorithms on multi-core CPUs and GPUs are considered. Issues related to distributed systems will be dealt with in a separate section later in this chapter.

3.2.1   Parallelism

The parallel graph algorithms presented in this chapter parallelize the outermost loop (these do not exploit nested parallelism). That is, the outermost `foreach` statement is run in `parallel` mode, and any nested `foreach` statements are run in sequential mode. This is quite common in both auto-parallelizing compilers as well.Exploiting nested parallelism requires efficient scheduling of iterations that balance work among threads. Parallelization of regular nested loops is based on thread and block based mapping. In this case, the number of iterations of an inner loop does not vary across the iterations of the outer loop. However, irregular nested loops do not benefit from this simple strategy. In the case of irregular nested loops, the use of thread-based mapping on the outer-loop may cause warp divergence (i.e., different threads are assigned different amounts of work), while the use of block-based mapping will lead to uneven block utilization, which in turn may cause GPU under-utilization.

3.2.2   Atomicity

The execution pattern of a graph algorithm may vary for graphs of even the same size ($|V|$ and $|E|$). Prediction of the runtime control flow and execution time from the size of the graph is impossible. In a parallel implementation of a graph algorithm,

Table 3.2 Atomic operations

Operation	CPU/GPU	Inputs	Operation	Return
CAS	Yes/Yes	x, y	if (mem.val==x) then mem.val = y;	mem.oldval
ADD	Yes/Yes	y	mem.val = mem.val+ y;	mem.oldval
MIN	Yes/Yes	y	if (mem.val > y) mem.val=y;	mem.oldval

two or more threads may try to update the same memory location at the same time. Such a race condition is handled by *atomic operations* provided by the hardware, which create *critical sections* that are executed by only one thread at a time. A few variants of the atomic operations relevant to graph algorithms are shown in Table 3.2. The `atomic` language construct used to create such critical sections[1] are implemented using atomic operations provided by the hardware. The atomic operations listed in the table are typically supported by both modern CPU and GPU. All the atomic operations return the value stored in the memory location before the operation was initiated.

3.2.3 An Example

Figure 3.1 shows a sample directed graph with four vertices and five edges. Figure 3.2 shows one of the possible parallel execution patterns of the SSSP computation in which, in each iteration, all the vertices are processed in parallel on a multi-core processor. A parallel version of the SSSP algorithm is discussed in detail in a later section in Algorithm 3.5. Assume that the outermost `forall` loop (Line 10 in Algorithm 3.5) is run in parallel. The vertex V_0 is considered as the *source* vertex. The execution sequence depends on the way threads are scheduled on the cores in addition to the graph topology. The *distance* of the source vertex is initialized to zero. The value is set to infinity (∞) for all the other vertices. The *distance* is reduced in each iteration until a fixpoint is reached (see Fig. 3.2). The two edges $V_1 \rightarrow V_2$ and $V_3 \rightarrow V_2$ have the same sink vertex V_2 and different source vertices (V_1 and V_3). When the two source vertices try to reduce the *distance* of the vertex V_2 at the same time using two different threads, there is a possible *race condition*. Assuming that both the vertices have passed the condition for update, V_3 may write 175 into V_2 followed by V_1 again "updating" V_2 to 300 (wrong distance)! A parallel implementation should use an atomic operation provided by the programming language to avoid such race conditions. The computation may be incorrect if atomic operations are not used in parallel implementations.[2]

---

[1] These are used in the forthcoming sections of this chapter.

[2] In this example, one more iteration of updates sets the mistake right. This is in general true only for *monotonic* and topology-driven algorithms [53].

Fig. 3.1 A directed input graph

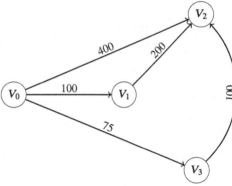

Fig. 3.2 Execution of SSSP computation on the graph object of Fig. 3.1

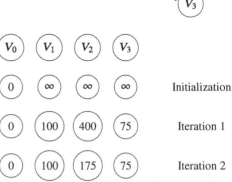

3.2.4   Graph Topology and Classification

Graphs in the real world vary in different aspects. Parallel implementations of graph algorithms do not always perform very well for all types of graphs and on all platforms. For example, road network graphs have large diameters, and the amount of parallelism available in them is less for *topology-driven* algorithms (e.g., BFS, see Algorithm 3.3). For road networks, such algorithms may perform very poorly on GPUs, but may perform reasonably on multi-core CPUs. Social network graphs follow the *power-law* degree distribution, and the difference in the maximum and the minimum degrees of the graph is very high. *Data-driven* algorithms such as minimum spanning tree computation perform well on GPUs and CPUs for many types of graphs, but not for social network graphs.

Random graphs follow uniform degree distribution and have a low diameter, and many parallel algorithms of both topology-driven and data-driven varieties perform well for these graphs on both GPUs and multi-core CPUs. The type of algorithm (topology-driven or data-driven) and the type of graph must be borne in mind while implementing parallel algorithms.

3.2.5 Push vs. Pull Computation

Graph algorithms are more efficient when synchronization and communication overheads are reduced in addition to a coalesced memory access pattern on GPUs. Push and pull versions of graph algorithms have been explored in [39, 54]. We explain the push and pull computation models with the pagerank algorithm as an example.[3]

Pull-based pagerank computation is shown in Algorithm 3.1. Each vertex p reads (pulls) the *pagerank* value of all the neighbouring vertices t with incoming edges $e : t \rightarrow p \in E$ (see Lines 4–8), and updates its pagerank. The parallel implementation does not require an `atomic` section as each thread updates (writes) its own pagerank value with only a *read* operation on the neighbouring vertices. Another example of a pull-based computation is in SSSP computation. Each thread operating on a vertex may pull the distance via its incoming neighbors (alternative distance of a node = distance of the incoming neighbor + weight on the edge between the neighbor and the node). Since each node's distance is being written to by maximum one thread, no atomics are necessary while updating the distance.

Algorithm 3.1: Parallel Pagerank—Pull Based

```
 1  d = 0.85; // damping factor
 2  pagerank(Point p, Graph G) {
 3      val = 0.0;
 4      foreach( Point t In p.innbrs ){
 5          if( t.outdegree > 0 ){
 6              val += t.pr ÷ t.outdegree;
 7          }
 8      }
 9      p.pr=val× d + (1 - d) ÷|G.V|;
10  }
11  computePR(Graph G, MAX_ITRS) {
12      foreach( t In G.V) in parallel t.pr = 1 ÷|G.V|;
13      itr = 1;
14      while( itr <MAX_ITRS ){
15          foreach( Points p In G ) in parallel {
16              pagerank(p, G);
17          }
18          itr++;
19      }
20  }
```

A *push* based computation of pagerank is shown in Algorithm 3.2. In this implementation, each vertex p updates (pushes) the pagerank of all the neighbouring

---

[3]Damping factor in Algorithms 3.1 and 3.2 is the probability at each page, that a "random surfer" will get bored and request another random page.

vertices t with an outgoing edge $p \rightarrow t \in E$. The update operation must be atomic, as two threads (vertices) p_1 and p_2 may try to update the pagerank of the common vertex t using two different edges $p_1 \rightarrow t$ and $p_2 \rightarrow t$ (see Line 5, which is enclosed in an atomic section). Push-based algorithms may be slower than pull-based algorithms, due to the presence of atomic operations. However, as demonstrated in [39], appropriate scheduling may make a push-based algorithm faster than a pull-based algorithm. Algorithm 3.3 (BFS) is another example of push-based computation. Each vertex p updates (pushes) attribute *dist* of all the neighbouring vertices t with an outgoing edge $p \rightarrow t \in E$. The update operation is atomic.

Push and pull models can be combined to arrive at a hybrid *pull-push-based* model, which is also used for the *pagerank* algorithm [39]. In this model, a vertex reads values from its in-neighbours (pull), and also updates value of its out-neighbours (push). The pull-push-based model of computation may look more expensive than pull and push models due to the presence of many more read and write operations. But pull-push model of execution results in faster information flow in a single iteration and appropriate scheduling delivers better results than a pure pull-based algorithm.

Algorithm 3.2: Parallel Pagerank: Push Based

```
 1  d = 0.85; // damping factor
 2  Update_Val(Point p, Graph G) {
 3      foreach( Point t In p.outnbrs ) in parallel {
 4          atomic
 5          |   t.val += p.pr ÷ p.outdegree;
 6          end
 7      }
 8  }
 9  Update_Pagerank(Point p, Graph G) {
10      |   p.pr = val × d + (1 - d) ÷ |G.V|; p.val = 0;
11  }
12  computePR(Graph G, MAX_ITR) {
13      foreach( t In G.V ) in parallel {
14          |   t.pr = 1 ÷ |G.V|; t.val = 0;
15      }
16      itr=0;
17      while( itr <MAX_ITR ){
18          foreach( Points p In G ) in parallel {
19              |   Update_Val(p, G);
20          }
21          foreach( Points p In G ) in parallel {
22              |   Update_Pagerank(p, G);
23          }
24          itr++;
25      }
26  }
```

3.2.6 Topology Driven vs. Data Driven

Parallel graph algorithms can be viewed as an iterated application of an abstract *operator* to *active* nodes in the graph. As an example, the abstract operator in pull-based *pagerank* algorithm is the updating of pagerank after reading the pagerank values of the neighbours of a vertex in a graph. *Activity* (performed through an operator) is generally local to a vertex, and affects only a small neighbourhood of the vertex. Therefore, several non-overlapping activities may proceed in parallel.

Topology-driven and *Data-driven* implementations of parallel algorithms are common. All the nodes[4] are considered as *active* in a topology-driven implementation. This leads to the application of the operator at all nodes, even if there is nothing to do at some nodes. For example, in BFS (see Algorithm 3.3), only the source node does some work in the first iteration and all other nodes have nothing to do. In the second iteration, all nodes at a distance of one from the source node have work to do and the others have nothing to do, and so on. However, in a topology-drive implementation, threads would be assigned to each node irrespective of whether work needs to be done at the node or not. A topology-driven variant is easier to implement than data-driven ones, specially on GPUs. GPUs have a huge number of cores and they can create a large number of threads very cheaply. Each thread needs to perform only a small amount of work corresponding to the operator, and programming such kernels (threads) is easy. However, efficiency of implementation does suffer, with many cores either doing nothing or doing redundant work.

In contrast, since multi-core CPUs have much fewer cores than GPUs, they often do not benefit from such implementations. Of special mention are very large and sparse graphs, on which topology-driven implementations on multi-cores perform poorly, due to relatively smaller number of edges (compared to the number of nodes).

Data-driven implementations apply the operator only at nodes that may have some work to do, and they do this by maintaining a queue of such active nodes. Each thread removes a node from the queue and applies the operator. Due to this action, some nodes that were inactive so far, may become active and will need to be inserted into the queue. Such parallel insertions and deletions on the queue (by the threads) need appropriate synchronization in order to prevent them from interfering with each other. Data-driven implementations are the rule as far as multi-core CPUs are concerned. However, such implementations on GPUs are far more challenging due to queue management and synchronization. It would be ideal if a library or a language provides a concurrent queue, and relieves programmers to focus on the algorithm development, rather than the clumsy details of the underlying synchronization mechanism.

---

[4]Vertices are also referred to as nodes.

3.2.6.1   A Simple Topology-Driven Breadth-First Search Algorithm

Algorithm 3.3 shows the pseudo code for topology-driven parallel breadth-first search on a (directed or undirected) graph. Breadth-first traversal repeatedly visits all the nodes (in parallel, one thread per node) in an increasing order of distance (i.e., number of edges from the source vertex). The `if` program block (Lines 11–15, Algorithm 3.3) updates the distance of the vertex t for an edge $p \rightarrow t \in E$.

Algorithm 3.3: Parallel Breadth-First Search (Topology-Driven)

```
1  TD_BFS(Point src, Graph G) {
2      foreach( Point p In G ) in parallel {
3          p.dist = ∞;
4          p.pred = NULL;
5      }
6      src.dist = 0;
7      while( 1 ){
8          changed = False;
9          foreach( Point p In G ) in parallel {
10             foreach( Point t In p.outnbrs ){
11                 atomic if ( t.dist >(p.dist + 1) ){
12                     t.dist = p.dist + 1;
13                     t.pred = p;
14                     changed = True;
15                 }
16             }
17         }
18         if(changed == False) break;
19     }
20 }
```

The `if` block needs to be atomic to avoid race conditions. For example, consider the graph in Fig. 3.1. Ignore the weights on the edges. Let the source node be V_0, and assume that V_0 first updates V_1 and then V_0 and V_1 race to update V_2. Without an atomic block, it is possible that both V_0 and V_1 pass the test in the `if` condition, V_0 first updates V_2 to 1 and then V_1 updates V_2 to 2! This is clearly wrong. With an atomic block, if V_0 passes the test first, it also gets to update V_2 first and then V_1 fails the test. However, if V_2 passes the test first and updates V_2 to 2, V_0 will still update V_2 to 1 when it gets its chance.

In the worst case, the number of iterations of the algorithm equals the maximum BFS distance value. In the initial iteration, the BFS distance of the neighbours of the source vertex is reduced from ∞ to one. The running time of Algorithm 3.3 on *road networks* is higher than that on *social networks* or *random* graphs of the same size, because *road network* graphs have a high *diameter*. A lock-free (no atomic block) version of BFS is presented in Algorithm 5.5 in Sect. 5.4.1.

3.2.6.2 A Simple Data-Driven Breadth-First Search Computation

Algorithm 3.4 presents a data-driven variant of BFS, which has better work-efficiency compared to its topology-driven counterpart. It works with a variable number of threads, T, and therefore scales better than the topology-driven algorithm as the number of vertices increases. The number of threads ($num\_threads$) is fixed at the beginning of the execution and does not change throughout the execution of the algorithm. For a graph $G = (V, E)$, V is partitioned into disjoint subsets such that

$$\bigcup_{i=0}^{num\_threads-1} V_i = V$$

BFS levels of vertices are recorded in the array bfs_level. Each thread numbered i exclusively owns partition V_i, and handles the queue Q_i. The queues Q_i (i $=$ $0, \ldots,$ num_threads-1) store the *frontier* of vertices as BFS progresses. All these queues are initialized to ϕ, except the one corresponding to k, where V_k contains the node src, the source node form which BFS begins (see loop at Line 6).

The loop at Line 17 advances the frontier of BFS by collecting and distributing the nodes adjacent to each node removed from the various queues owned by the threads. Among the collected nodes by thread i, nodes present in the partition V_j are pushed into the queue $Q_{i,j}$ (see the loops at Lines 22 and 25). The loop at Line 34 forms the new queue for each thread Q_i, and assigns the BFS level to each node inserted into the queue Q_i (see Line 39). The enclosing while loop at Line 16 exits only when all the queues Q_i are empty simultaneously.

It is easy to see that this BFS algorithm is data-driven; there are active queues (Q_i) and threads operate on these queues. By using an appropriate number of threads, the number of threads that are inactive can be reduced to acceptable levels. An improved version of this algorithm that is adaptive in the number of threads is presented in [55].

3.3 Different Ways of Solving a Problem

There are many possible implementations for a given problem. As an example, different ways of computing SSSP (using different parallel algorithms) for a directed graph $G(V, E)$ are discussed. Figure 3.3 shows a sample weighted directed graph. Vertex s is identified as the source vertex. The SSSP algorithms discussed here terminate only if there is no negative cycle in the graph. While having this graph here makes the subsection independent, we could have utilized a single graph for all SSSP computations.

Algorithm 3.4: Parallel Breadth-First Search (Data-Driven)

```
 1  DD_BFS(Point src, Graph G(V,E), num_thread) {
 2      Partition V into disjoint subsets V_0, V_1, ..., V_{num_thread-1};
 3      level = 1;
 4      flag = 1; unmark all vertices;
 5      bfs_level[src] = 0; // One entry for each vertex in bfs_level
 6      foreach( i ∈ {0, ..., num_threads − 1} ) in parallel {
 7          stop[i] = 0;
 8          if ( src ∈ V_i ){
 9              |  Q_i = {src};
10          }
11          else{
12              |  Q_i = ϕ;
13          }
14      }
15      // Loop until Q_0, Q_1, ..., Q_{num_thread-1} are all ϕ simultaneously.
16      while( flag == 1 ){
17          foreach( i ∈ {0, ..., i − 1} ) in parallel {
18              if ( Q_i ≠ ϕ ){
19                  Neighbours_i = ϕ;
20                  foreach( Point v In Q_i ){
21                      Q_i = Q_i − {v};
22                      foreach( Point n In v.outnbrs ){
23                          |  Neighbours_i = Neighbours_i ∪ {n};
24                      }
25                      foreach( j ∈ {0, ..., num_threads − 1} ){
26                          |  Q_{i,j} = Neighbours_i ∩ V_j;
27                      }
28                  }
29              }
30              else{
31                  |  stop[i] = 1;
32              }
33          }
34          foreach( i ∈ {0, ..., i − 1} ) in parallel {
35              R_i = ϕ;
36              foreach( j ∈ {0, ..., i − 1} ){
37                  |  R_i = R_i ∪ Q_{i,j};
38              }
39              foreach( unmarked v ∈ R_i ){
40                  mark v; Q_i = Q_i ∪ {v};
41                  bfs_level[v] = level;
42              }
43              if ( Q_i ≠ ϕ ){
44                  |  stop[i] = 0;
45              }
46          }
47          level = level + 1; flag = 0;
48          foreach( i ∈ {0, ..., i − 1} ) in parallel {
49              if ( stop[i] == 0 ){
50                  |  flag = 1;
51              }
52          }
53      }
54  }
```

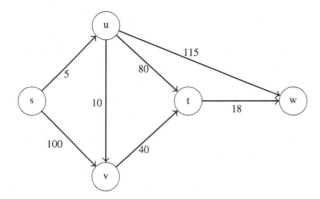

Fig. 3.3 An input graph for SSSP algorithm

Table 3.3 SSSP: vertex-based computation (Algorithm 3.5)

Vertex (dist, pred)	s	u	v	t	w
Initial	(0, −)	(∞, −)	(∞, −)	(∞, −)	(∞, −)
itr1	(0, −)	**(5,s)**	**(100,s)**	(∞, −)	(∞, −)
itr2	(0, −)	(5,s)	**(15,u)**	**(85,u)**	**(120,u)**
itr3	(0, −)	(5,s)	(15,u)	**(55,v)**	**(103,t)**
itr4	(0, −)	(5,s)	(15,u)	(55,v)	**(73,t)**
Final	(0, −)	(5,s)	(15,u)	(55,v)	(73,t)

3.3.1   Vertex-Based SSSP Computation

Algorithm 3.5 is a topology-driven, push-based, and vertex-based SSSP computation. *p.dist* stores the current shortest distance from the source *s* to *p*, and *p.pred* stores the predecessor of *p* in such a shortest path. Table 3.3 shows a possible SSSP computation on the graph object using Algorithm 3.5. The table shows the vertices whose distances are reduced in bold face. It shows the predecessor in the shortest path along with the shortest distance. Lines 2–6 of Algorithm 3.5 initialize *dist* and *pred*. The SSSP computation happens in Lines 7–19. The outer while loop iterates until each vertex reachable from the source vertex gets a stabilized *dist* value. In the parallel loop, each vertex *p* updates the *dist* value of each of its neighbours (consider one, say *t*) if going to *t* through *p* is cheaper than current path to *t*. It is important to note that this algorithm computes shortest distances correctly irrespective of the order of updates in the parallel loop and that an atomic block is necessary for the update (see example the discussion in Sect. 3.2.3).

Algorithm 3.5: Parallel Vertex Based SSSP Computation

```
1  SSSP(Graph G) {
2      foreach( Point p In G ) in parallel {
3          p.dist = ∞;
4          p.pred = NULL;
5      }
6      src.dist = 0;
7      while( True ){
8          changed = False;
9          foreach( Point p In G ) in parallel {
10             foreach( Point t In p.outnbrs ){
11                 atomic if ( t.dist >(p.dist + G.getWeight(p, t)) ){
12                     t.dist = p.dist + G.getWeight(p, t);
13                     t.pred = p;
14                     changed = True;
15                 }
16             }
17         }
18         if(changed == False)break;
19     }
20 }
```

Algorithm 3.6: Parallel Worklist Based SSSP Algorithm

```
1  SSSP(Graph G) {
2      foreach( each vertex v in V ) in parallel {
3          distance[v] = ∞;
4          predecessor[v] = NULL;
5      }
6      distance[s] = 0;
7      Worklist<vertex>current, next;
8      current.add(s);
9      while( True ){
10         foreach( each vertex u In current ) in parallel {
11             for each edge (u, v) with weight w {
12                 atomic if ( distance[u] + w <distance[v] ){
13                     distance[v] = distance[u] + w;
14                     predecessor[v] = u;
15                     next.add(v);
16                 }
17             }
18         }
19         swap(current, next);
20         ifcurrent.size == 0 break;
21     }
22     return distance[], predecessor[];
23 }
```

3.3.2 Worklist-Based SSSP Computation

Algorithm 3.6 shows the pseudo code for a worklist based SSSP computation. This is a data-driven implementation. The variables, *distance* and *predecessor* of all the vertices are initialized as in Algorithm 3.5 and *distance* of the source vertex s is then reset to zero. Computation proceeds using two worklists, *current* and *next*, which store a subset of the vertices of the graph. The source vertex s is added to the worklist *current* (Line 8). The computation takes place in the while loop (Line 9–21), which exits when the shortest distances to all the *reachable* vertices have been computed. In the while loop, each vertex u in the worklist *current* is considered (Line 10) and its outgoing edges are processed in the inner loop (Lines 11–17). For each edge $u \rightarrow v \in E$ considered for processing, the distance reduction is done as in Algorithm 3.5. If the distance of the vertex v is reduced, it is added to the worklist *next* (Line 15), which contains the vertices to be processed in the next iteration of the while loop. Once all the elements in *current* are processed, worklists *current* and *next* are swapped (Line 19), making the *size* of *next* as zero (empty worklist). After the swap of *next* and *current* worklists, size of *current* will be number of elements added to *next* during the last execution of the loop in Lines 10–18. If no elements are added to *next*, size of *current* after swap operation will be zero, this is the fixpoint for computation and the algorithm terminates. The above algorithm has a computational complexity of $O(V + E)$. Table 3.4 shows the iterations of Algorithm 3.6 for the input graph in Fig. 3.3.

3.3.3 Edge-Based SSSP Computation

Vertex-based processing assigns a vertex (or a group of vertices) to each thread. Each thread then typically iterates over the assigned vertex's neighbors to apply the computation operator. For graphs having high variance in degree distribution, the neighborhood sizes of vertices may vary considerably (e.g., in social networks). This leads to load-imbalance as the thread finishing its operation early needs to wait for a thread assigned to high-degree vertex. To address this issue, threads may

Table 3.4 SSSP—worklist based computation (Algorithm 3.6)

Iteration	Active vertices	$s(dist, pred)$	$u(dist, pred)$	$v(dist, pred)$	$t(dist, pred)$	$w(dist, pred)$
Initial	NIL	$(0, -)$	$(\infty, -)$	$(\infty, -)$	$(\infty, -)$	$(\infty, -)$
itr1	s	$(0, -)$	**(5,s)**	**(100,s)**	$(\infty, -)$	$(\infty, -)$
itr2	v, u	$(0, -)$	(5,s)	**(15,u)**	**(85,u)**	**(120,u)**
itr3	w, t, v	$(0, -)$	(5,s)	(15,u)	**(55,v)**	**(103,t)**
itr4	w, t	$(0, -)$	(5,s)	(15,u)	(55,v)	**(73,t)**
itr5	NIL	$(0, -)$	(5,s)	(15,u)	(55,v)	**(73,t)**

Algorithm 3.7: Parallel Edge Based SSSP Computation

```
1  SSSP(Graph G) {
2      foreach( Point p In G ) in parallel {
3          p.dist = ∞;
4          p.pred = NULL;
5      }
6      src.dist = 0;
7      while( True ){
8          changed = False;
9          foreach( edge p → t In G ) in parallel {
10             atomic if ( t.dist >(p.dist + G.getWeight(p, t)) ){
11                 t.dist = p.dist + G.getWeight(p, t);
12                 t.pred = p;
13                 changed = True;
14             }
15         }
16         if(changed == False)break;
17     }
18 }
```

---

be assigned to edges (or a group of edges), leading to near-perfect load-balance. Algorithm 3.7 performs such a parallel edge based computation of SSSP. The major advantage of Algorithm 3.7 over Algorithm 3.5 is that there is only one parallel loop which processes all the edges. Edge based processing is efficient in GPU for social network graphs which have a high variance in degree distribution, because it results in less *warp divergence*.

3.3.4 Δ-Stepping SSSP

A topology-driven processing is suitable for low-diameter graphs. In contrast, a data-driven processing works well for low-degree graphs. A data-driven processing crucially relies on the fact that the vertices are processed largely in the increasing order of their distances (in an extreme case, such a processing resembles work-efficient Dijkstra's algorithm). Road networks defy both these kinds of processing, since they have high-diameter. Thus, a topology-driven processing leads to too much extra work since only a few vertices get active in each iteration due to low-diameter; whereas the amount of work done per iteration in a data-driven approach is also very small leading to very slow progress and low concurrent execution. Due to high diameter, the number of iterations can be very high (e.g., California road network has a diameter of 849 requiring at least those many iterations). A generalization of these two extremes is Δ-Stepping, wherein vertices are ordered based on distance values leading to quicker fixpoint computation, but also some redundant computation is allowed to improve concurrency.

The Δ-Stepping SSSP performs well on a multi-core CPU, especially on road networks. The parallelism available is not sufficient to engage all the cores on a GPU, and thereby, the performance is not good on GPU devices. In the Δ-Stepping SSSP Algorithm [10], vertices are ordered using a set of *worklists* called *buckets* representing priority ranges of Δ, where Δ is a positive value. The *bucket* $B[i]$ will have vertices whose current *distance* value is given by $(i - 1) \times \Delta \leq distance < i \times \Delta$. The *buckets* are processed in an increasing order of index value i and a bucket $B[i]$ is processed only after bucket $B[i - 1]$ is processed. Algorithm 3.8 shows the Δ-Stepping SSSP algorithm.

The function *Relax* takes as argument a vertex v and an integer value x. If the current *distance* of v is greater than x, the vertex v is removed from the current bucket $B[distance(v) \div \Delta]$ and it is added to the bucket $B[x \div \Delta]$. Then the *distance* of vertex v is reduced to x (Line 5).

Algorithm 3.8: Δ-Stepping SSSP Algorithm

```
1  Relax( vertex v, int x) {
2      if( (x <distance(v) ) ){
3          Bucket B[ distance(v)÷ Δ] = B[distance(v)÷ Δ] \ v;
4          Bucket B[ x÷ Δ] = B[x÷ Δ] ∪ v;
5          distance(v) = x;
6      }
7  }
8  SSSP (Graph G) {
9      foreach( each v in V ) in parallel {
10         Set heavy(v) = { (v, w) ∈ E: weight(v, w) >Δ }
11         Set light (v) = { (v, w) ∈ E: weight(v, w) <= Δ }
12         distance (v) = INF; // Unreached
13     }
14     relax(s, 0); // bucket zero will have source s.
15     i = 0;
16     // Source vertex at distance 0
17     while( NOT isEmpty(B) ){
18         Bucket S = φ;
19         while( B[i] ≠ φ ){
20             Set Req = {(w, distance(v) + weight(v, w)) : v ∈ B [i] ∧ (v, w) ∈ light(v)};
21             S = S ∪ B[i];
22             B[i] = φ;
23             foreach ((v, x) ∈ Req) in parallel Relax(v, x);
24         }
25         //done with B[i]. add heavy weight edge for relaxation
26         Req = { (w, distance(v) + weight(v, w)): v ∈ S ∧ (v, w) ∈ heavy(v) }
27         foreach((v, x) ∈ Req)in parallel relax(v, x);
28         i = i + 1
29     }
30 }
```

The *SSSP* algorithm works in the following way. Initially for each vertex v in the graph, two sets *heavy* and *light* are computed, where:

$heavy(v) = (\ \forall\ (v, w) \in E) \bigwedge (weight(v, w) > \Delta\ )$

$light(v) = (\forall\ (v,w) \in E) \bigwedge (weight(v, w) \leq \Delta)$

Then the *distance* of all the vertices is made ∞ (Lines 9–13). The core of the algorithm starts by relaxing the *distance* value of source vertex s in Line 14 with a *distance* value of zero. This adds the source vertex to *bucket* zero (Line 21). Then the algorithm enters the while loop in Lines 17–29, processing *buckets* in an increasing order of index value i, starting from zero.

An important feature of the algorithm is that, once the processing of *bucket* $B[i]$ is over, no more elements will be added to the bucket $B[i]$, when the buckets are processed with increasing values of index i. A *bucket* $B[i]$ is processed in the while loop (Lines 19–24). At first, for all vertex v in *bucket* $B[i]$, all edges $v \to w$ $\in light(v)$ are considered and the pair $(w, distance(v) + weight(v, w))$ is added to the Set *Req*. This is followed by the Set S added with all the elements in $B[i]$ (Line 21) and $B[i]$ made empty (Line 22). Then *Relax()* function is called for all elements in Set *Req*. This adds new elements to multiple buckets. It can add a vertex w to a bucker $B[k]$ where $k \geq i$. The bucket to which the vertex w is added is $(dist[v]+weight(v,w)) \div \Delta$.

The vertex w is added to the bucket $B[i]$ if $distance[w] \geq i \times \Delta$ and there can be element $(w, x) \in Req$ where $x < i \times \Delta$. Here $x = distance(v) + weight(v, w)$ for an edge $v \to w$. Once the *bucket* $B[i]$ becomes empty after a few iterations, the program exits the while loop (Lines 19–24). Now all edges $v \to w \in heavy(v)$ are considered and the pair $(w, distance(v) + weight(v, w))$ is added to the Set *Req*(Line 26). The edges in Set heavy have weight $> \Delta$ and this makes $\forall(v, x) \in Req\ x > i \times \Delta$. So new elements will be added to bucket $B[j]$ when the Set *Req* is relaxed where $j > i$ (Line 27). Now value of i is incremented by one (Line 28) and algorithm starts processing *bucket* $B[i + 1]$. Algorithm terminates when all the *buckets* $B[i]$, $i \geq 0$ are empty. The performance of the algorithm depends on the input graph and the value of the parameter Δ, which is a positive value. For a Graph $G(V, E)$ with random edge weights, maximum node degree d ($0 < d \leq 1$), the sequential Δ-stepping algorithm has a time complexity of $O(|V| + |E| + d \times P)$, where P is the maximum SSSP distance of the graph. So, this algorithm has running time which is linear in $|V|$ and $|E|$.

3.4   A Note on Efficient Parallel Implementations of Graph Algorithms

While graph algorithms, in general, exhibit enough parallelism, some algorithms are notoriously difficult to parallelize. The depth first search (DFS) algorithm is inherently sequential [56]. The strongly connected components (SCC) algorithm discussed in Algorithm 2.6 of the previous chapter is difficult to parallelize as DFS is used. Algorithm 2.7 which uses a divide-and-conquer strategy is more suitable for parallelization.

The BFS algorithm by Merrill et al. [57] performs well on single and multi-GPU devices. The algorithm relies on prefix-sum computation and task management computed therefrom. The algorithm uses duplicate detection techniques to avoid race condition. The algorithm achieves high performance on different graphs which vary in topology. The Δ-Stepping algorithm for SSSP computation is very efficient on multicore-CPU. The amount of parallelism available in the algorithm is not sufficient to use all the GPU cores. The Δ-Stepping algorithm is much faster for graphs with a large *diameter* such as road-networks, when executed on multi-core CPUs.

Large scale graph processing on a single machine is possible by using disk storage. Large graphs are processed on a single machine, with the graph being split into smaller subgraphs on a disk. The subgraphs are loaded one by one into RAM and then processed. Such processing is called *out-of-core* processing. When a cluster of machines is available, the graph may be partitioned across various machines in the cluster [58].

3.5   Some Important Parallel Graph Algorithms

In the last few sections, several parallel algorithms were presented. The following sections present a few more examples of important parallel graph algorithms.

3.5.1   Parallel Computation of Maximal Independent Sets

This is similar to the sequential MIS algorithm presented in Sect. 2.2.8. It is a randomized parallel algorithm. The set S accumulates the MIS. To begin with, a randomized value is attached to each vertex as a label, and isolated vertices are added to S (Line 4) using the parallel foreach statement. Line 10 is the an important step. It chooses a vertex v for inclusion in S only if it has a label value lesser than all its neighbours. All the neighbours of v and edges incident on v are removed from the graph (Line 18) and the reduced graph is again subject to the same process (Line 25). The algorithm terminates when no further vertices are left to process. This randomized algorithm terminates in $O(\log n)$ time with high probability. The parallel operations of set union and set difference are usually available in most graph processing frameworks.

Algorithm 3.9: Parallel Maximal Independent Set Computation

```
 1  Maximal_Independent_Set(Graph G(V, E)) {
 2      if ( V ≠ φ ){
 3          Set S = φ, C = V, D = E;
 4          foreach( Vertex v In V ) in parallel {
 5              v.label = rand(); // Each v picks a random value in [0:1].
 6              if ( v.nbrs == φ ){
 7                  S = S ∪ {v}; // Add v to S if it is an isolated vertex.
 8              }
 9          }
10          foreach( Vertex v In V ) in parallel {
11              flag = 0;
12              foreach( Vertex u In v.nbrs ){
13                  if ( u.label < v.label ){
14                      flag = 1;
15                      break;
16                  }
17              }
18              if ( flag == 0 ){
19                  S = S ∪ {v}; C = C \ ({v} ∪ v.nbrs);
20                  foreach( Vertex u In v.nbrs ){
21                      D = D \ {(v, u)};
22                  }
23              }
24          }
25          return (S ∪ Maximal_Independent_Set(G(C,D)));
26      }
27      else{
28          return φ;
29      }
30  }
```

3.5.2 Parallel Computation of Strongly Connected Components

Parallel computation of SCCs is based on the sequential version presented in Sect. 2.7. Since the three partitions of vertices on which SCC is recursively called are all disjoint, the recursive calls can be processed in parallel. To begin with, in each call to SCC, vertices which have only incoming or outgoing arcs are single vertex SCCs on their own, and therefore can be processed early. This process is called trimming [59]. This can be done in parallel for all vertices, but has to be repeated several times because trimming gives rise to other vertices that can be trimmed (as in the case of a chain of edges in the same direction). Trimming avoids spending much time on single vertex SCCs. A more efficient implementation which ensures work balance among threads, specially for real-world graphs, is reported in [60].

Algorithm 3.10: Parallel Algorithm for Finding Strongly Connected Components

---

1 *Parallel_SCC(Graph G(V, E), Vertexset P, Vertexsetcollection StrongCompSet)* {
2   // P is the vertex set on which forward and backward searches are performed.
3   // When SCC is called for the first time, $P = V$. Then onwards, $P \subset V$. //
     StrongCompSet is the collection of strongly connected components,
4   // represented by their respective vertex sets.
5   // Now trim vertices which have only incoming or outgoing arcs.
6   **while(** $True$ **){**
7   changed = False;
8   **foreach(** *Vertex v In P* **) in parallel** {
9   **if(** *(v.innbrs == ϕ)$\vee$(v.outnbrs == ϕ)* **){**
10                $StrongCompSet = StrongCompSet \cup \{v\}$;
11                $P = P \setminus \{v\}$;
12                $changed$ = True;
13   }
14   }
15   if(changed=False) break;
16   }
17   **if** $(P == \phi)$ **return**;
18   **Select** v uniformly at random from P;
19   // v is the pivot vertex from which forward and backward searches are made.
20   **begin parallel**
21   V_{fwd} = Fwd-Reachable(G, P, v);
22   // This is a parallel BFS on G, starting from v. Vertex set for BFS is limited to P.
23   V_{bwd} = Bwd-Reachable(G^T, P, v);
24   // This is a parallel BFS on G^T, starting from v. Vertex set for BFS is limited to P.
25   **end**
26     $scc = V_{fwd} \cap V_{bwd}$;
27     $StrongCompSet = StrongCompSet \cup scc$;
28   //Parallel recursive calls to SCC
29   **begin parallel**
30   SCC($G(V, E), V_{fwd} \setminus scc, StrongCompSet$);
31   SCC($G(V, E), V_{bwd} \setminus scc, StrongCompSet$);
32   SCC($G(V, E), V \setminus (V_{fwd} \cup V_{bwd}), StrongCompSet$);
33   **end**
34 }

3.5.3 Parallel Computation of Connected Components

While connected components of an undirected graph can be found easily by a sequential DFS (or BFS) Algorithm 2.3, it is surprisingly much harder to do so using a parallel algorithm. DFS is not amenable to good parallelization as has been discussed in [61]. Even though BFS is parallelizable, poly-log time complexity cannot be guaranteed. Therefore, randomized parallel algorithms for finding connected components using BFS as a sub-algorithm have been proposed. This section describes one such approach from [62]. This is a randomized parallel algorithm that performs linear work and runs in $O(log^3 n)$ time with high probability, and is shown in Algorithms 3.11 and 3.12.

The algorithm partitions a graph into clusters, with each cluster being a tentative connected component (see Line 8, Algorithm 3.11), contracts the clusters into single nodes and creates a smaller graph after contraction (see Line 10, Algorithm 3.12), and repeats these steps until all the components are finalized. Function *Parallel_CC* (Line 39, Algorithm 3.12) recursively calls itself to perform these steps. The factor β which is a parameter to function *Decompose* is a parameter of an exponential probability distribution $((1/\beta)$ is the mean) and controls the diameter of clusters found by *Decompose*, and also inter-cluster edges. Larger values of β generate clusters of smaller diameter and hence will have more number of inter-cluster edges. Function *Decompose* is the heart of the algorithm.

In the foreach statement in Line 10 of function *Decompose*, a random number δ is generated for each vertex in parallel, based on an exponential distribution with a mean value of $(1/\beta)$. The foreach statement in Line 16 of function *Decompose* computes in parallel a *start* value for each vertex based on the value of δ_{max} computed in Line 13 of function *Decompose*. It is to be noted that *start* values may not be integers and there may be several vertices with *start* values in a given range, say x and $x + 1$, with x being an integer.

Line 19 of function *Decompose* begins the main loop of the function which iterates until all the vertices of the graph are visited ($O(log\ n)$ iterations will be performed). Each iteration begins by collecting all the vertices which are not yet visited ($v.compnum == \infty$) and whose *start* values are less than *round*+1 (*round* is initialized to zero), into a set called *Frontier* (see Line 20, Algorithm 3.11). The vertices in *Frontier* begin growing either the old cluster (to which they belong) or a new cluster (if they have just got into *Frontier*), in the loop at Line 27 of function *Decompose*, by taking one step, in which each vertex v looks at its neighbours in parallel. If the neighbour is still unvisited ($v.compnum == \infty$), and some other vertex has not annexed it already,[5] vertex v adds this neighbour to its own cluster by numbering the neighbour with its own component number. These new recruits are added to *NewFrontier* which becomes the *Frontier* for the next iteration with value *round*+1. After the clustering operation, *Decompose* returns a map of vertex labels and their component numbers, created by function *CreateCCmap*(see Line 1, Algorithm 3.11).

After calling *Decompose* once from function *Parallel_CC*, the clusters are contracted into single vertices in function *Contract* (see Line 10, Algorithm 3.12). New vertices and edges are created in the new graph and then the data structure for the new graph is created in Line 36[6] of Algorithm 3.12. If the new graph is

---

[5]This can happen due to concurrent processing of neighbours by vertices and hence requires a CAS instruction that ensures atomicity.

[6]Function *CreateNewGraph* is easy to write and depends on the data structure used to represent the graph. This is necessary to provide features to traverse neighbouring vertices and edges. Its details are not shown in the book.

not empty (see Line 42, Algorithm 3.12), function *Parallel_CC* is called again on the reduced new graph, and a new component map is collected. The old map and the new map are combined to yield a new map in the function *Relabel* (see Line 1, Algorithm 3.12), which is called towards the end of *Parallel_CC*. The combined new map is returned by *Parallel_CC*.

Algorithm 3.11: Functions Decompose and CreateCCmap Used in Function *Parallel_CC*

```
1  CreateCCmap(Graph G(V,E)) {
2      Collection CCmap = φ;
3      foreach( v In V ) in parallel {
4          │  CCmap = CCmap ∪ {(v.label, v.compnum)};
5      }
6      return CCmap;
7  }
8  Decompose(Graph G(V,E), float β) {
9      δ_max = 0; Collection Frontier = φ; numvisited = 0; round = 0;
10     foreach( v In V ) in parallel {
11         │  v.δ = exp_random(β);
12     }
13     foreach( v In V ) in parallel {
14         │  δ_max = max(δ_max, v.δ); // max is atomic.
15     }
16     foreach( v In V ) in parallel {
17         │  v.start = δ_max − v.δ; v.compnum = ∞;
18     }
19     while( numvisited < V.size ){
20         foreach( v In V ) in parallel {
21             if( v.start < round+1 ∧ v.compnum == ∞ ){
22                 Frontier = Frontier ∪ {v};
23                 v.compnum = v.label; // Vertices are assumed to have unique labels.
24             }
25         }
26         numvisited = numvisited + Frontier.size;
27         foreach( v In Frontier ) in parallel {
28             NewFrontier = φ;
29             foreach( i In v.nbrs ){
30                 if( i.compnum == ∞ ∧ CAS(i.compnum, ∞, v.compnum) ){
31                     │  NewFrontier = NewFrontier ∪ {i};
32                 }
33             }
34         }
35         round = round + 1;
36         Frontier = NewFrontier;
37     }
38     return CreateCCmap(G(V, E));
39 }
```

Algorithm 3.12: Parallel Algorithm for Finding Connected Components

---

```
1  Relabel(Collection CCmap1, Collection CCmap2) {
2      Collection CCmap3 = φ;
3      foreach( (v.label, v.compnum) In CCmap1 ) in parallel {
4          if( (v.compnum, v.compnumNew) ∈ CCmap2 ){
5              CCmap3 = CCmap3 ∪ {(v.label, v.compnumNew)};
6          }
7      }
8      return CCmap3;
9  }
10 Contract(Graph G(V, E)) {
11     V_new = φ; E_new = φ;
12     foreach( v In V ) in parallel {
13         v.newcreated = NULL;
14     }
15     foreach( v In V ) in parallel {
16         foreach( i In v.nbrs ){
17             if( i.compnum ≠ v.compnum ){
18                 if( v.newcreated == NULL ){
19                     v_new = CreateVertex(v); v_new.label = v.compnum;
20                     if( CAS(v.newcreated, NULL, v_new) ){
21                         V_new = V_new ∪ {v_new};
22                     }
23                 }
24                 if( i.newcreated == NULL ){
25                     i_new = CreateVertex(i); i_new.label = i.compnum;
26                     if( CAS(i.newcreated, NULL, i_new) ){
27                         V_new = V_new ∪ {i_new};
28                     }
29                 }
30                 if( (v_new, i_new) ∉ E_new ){
31                     E_new = E_new ∪ {(v_new, i_new)};
32                 }
33             }
34         }
35     }
36     G'(V', E') = CreateNewGraph(G(V, E), V_new, E_new);
37     return G'(V', E');
38 }
39 Parallel_CC(Graph G(V, E), float β) {
40     Collection CCmap1 = Decompose(G(V, E), β);
41     G_new(V_new, E_new) = Contract(G(V,E), CCmap1);
42     if( ( E_new == φ) ){
43         return CCmap1;
44     }
45     else{
46         CCmap2 = Parallel_CC(G_new(V_new, E_new), β);
47         CCmap3 = Relabel(CCmap1, CCmap2);
48         return CCmap3;
49     }
50 }
```

Figures 3.4a–d and 3.5a–i show how an undirected graph is processed to compute connected components using algorithm*Parallel_CC*. In Round $= 0$, vertices v_1 and v_4 become eligible (according to their *start* values which are not shown) and are placed in *Frontier*. They start inspecting their neighbours in parallel. v_1 grows the ball around it by including vertex v_2 into its cluster and both these vertices are given the same component number (*compnum*) of 1. Similarly, vertices v_3, v_4 and v_5 are given the same component number of 4. Edge (v_2, v_3) is now an inter-component edge. In Round $= 1$, vertices v_2, v_3, v_5, v_7 and v_9 become eligible based on their *start* values and start parallel inspection of their respective neighbours. v_2 and v_3 have no *unvisited* neighbours (see Fig. 3.4b). Vertices v_5 and v_7 have a common *unvisited* neighbour, v_6. Only one of them can include v_6 into their component and this choice being arbitrary, v_5 wins. Similarly, vertices v_7 and v_9 have a common *unvisited* vertex v_8, and in this case v_7 wins in the arbitrary choice (decided by the atomic CAS operation), and includes v_8 in its cluster. The clusters formed at the end of Round 1 are shown in Fig. 3.4c. The map of vertex labels and their component numbers (still tentative) are shown in Fig. 3.4d. No more rounds are possible and now the graph with four clusters is contracted.

The contracted graph is shown in Fig. 3.5a. It has four nodes corresponding to four clusters found so far. The process explained above is applied to this contracted graph and finally one component is found as shown in Fig. 3.5g, h. All the vertices of the original graph form a single component and the component number is 7 (it could have been any other label as well).

3.5.4 Parallel Computation of Betweenness Centrality (BC)

Section 2.3.5 presented a sequential algorithm to compute BC based on the Floyd-Warshall all-pairs shortest path algorithm. This was an $O(n^3)$ time algorithm. This section presents a parallel algorithm based on the more efficient algorithm due to Brandes [45]. The equation for computing BC in Eq. 2.1 may be rewritten as:

$$BC(v) = \sum_{s \neq v \neq t} \delta_{st}(v) \tag{3.1}$$

where, $\delta_{st}(v)$ is the *pairwise dependency* and is defined as below:

$$\delta_{st}(v) = \frac{\sigma_{st}(v)}{\sigma_{st}} \tag{3.2}$$

While findings shortest paths in unweighted undirected paths may be carried out by a cheaper BFS strategy, computing pairwise dependencies for each vertex is still

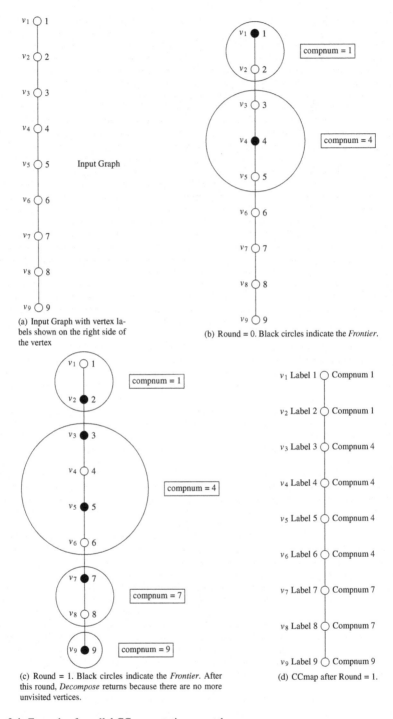

(a) Input Graph with vertex labels shown on the right side of the vertex

(b) Round = 0. Black circles indicate the *Frontier*.

(c) Round = 1. Black circles indicate the *Frontier*. After this round, *Decompose* returns because there are no more unvisited vertices.

(d) CCmap after Round = 1.

Fig. 3.4 Example of parallel CC computation—part 1

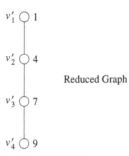

Reduced Graph

(a) Reduced Graph after contraction

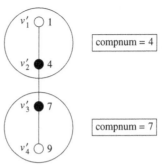

compnum = 4

compnum = 7

(b) Round=0. Black circles indicate the *Frontier*. After this round, *Decompose* returns because there are no more unvisited vertices.

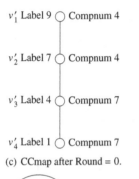

v'_1 Label 9 ◯ Compnum 4

v'_2 Label 7 ◯ Compnum 4

v'_3 Label 4 ◯ Compnum 7

v'_4 Label 1 ◯ Compnum 7

(c) CCmap after Round = 0.

$v_1"$ ◯ 4

Reduced Graph

$v_2"$ ◯ 7

(d) Reduced Graph after contraction

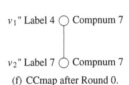

compnum = 7

(e) Round=0. Black circle indicates the *Frontier*. After this round, *Decompose* returns because there are no more unvisited vertices.

$v_1"$ Label 4 ◯ Compnum 7

$v_2"$ Label 7 ◯ Compnum 7

(f) CCmap after Round 0.

v'''_1 ◯ 7   Reduced Graph

(g) Reduced Graph after contraction. No more steps are possible.

v'''_1 Label 7 ◯ Compnum 7

(h) CCmap after no more steps are possible.

There is only one connected component. All the vertex labels map to component number 7.
(i) Final CCmap.

Fig. 3.5 Example of parallel CC computation—part 2

expensive. The computation of pairwise dependencies is carried out differently in Brandes' algorithm [45] using the following equations.

$$BC(v) = \sum_{s \neq v} \delta_s(v) \tag{3.3}$$

$$\delta_s(v) = \sum_{t \in V} \delta_{st}(v) \tag{3.4}$$

$$\delta_s(v) = \sum_{w:v \in pred(s,w)} \frac{\sigma_{sv}}{\sigma_{sw}}(1 + \delta_s(w)) \tag{3.5}$$

$$\sigma_{sv} = \sum_{u \in pred(s,v)} \sigma_{su} \tag{3.6}$$

The key to fast computation is the recurrence (3.5) that enables *independent* computation of fractional pairwise dependency contribution by each vertex after shortest path computation. The parallel algorithm now runs as follows. A BFS run concurrently from each vertex v of the graph computes the distance, prdecessors, and number of shortest paths (the arrays $v.distance[]$, $v.predecessor[]$ and $v.\sigma[]$, respectively) from v. Then the fractional contributions of each vertex to BC(v) are summed up to produce BC(v). The details are shown in Algorithm 3.13. This algorithm is based on [63]. Using fine grained locks and lock-free data structures as discussed in [63] yields even better performance.

Algorithm 3.13 processes each vertex of the graph sequentially, but the BFS and other operations in each iteration are parallel algorithms. It uses an array of stacks indexed by the BFS level to store the vertices visited at each level. This hierarchy is used to compute the *partial pairwise dependencies* in a bottom-up manner. After initializations (Lines 3–19), parallel BFS begins. Newly visited vertices are pushed onto the stack at the appropriate level (Line 24), and distance, predecessor and σ (number of shortest paths) values are updated (Line 31). The atomic sections are necessary since global variables *level, Stack_Array,* and σ are being manipulated by several concurrent operations. Operations beginning at Line 41 show the update of δ and BC values in a bottom-up manner using *Stack_Array* and *level*.

Figure 2.11a, b are reproduced in Fig. 3.6 for convenience. The trees produced by BFS initiated at each vertex along with partial δ values are shown in Fig. 3.7. Of course, tree structure is provided for better understanding and is not needed for the computation—level information is sufficient. Consider BC computation for vertex 1 in Figs. 3.6a and 3.7a. To begin with, vertex 1 is pushed onto Stack_Array[0]. Then, neighbours of vertex 1, vertices 6 and 3 are pushed onto Stack_Array[1], indicating that they would be processed after advance of BFS by one level. Vertex 1 is put on predecessor lists of both vertices 6 and 3. Distance and σ values of both vertices 6 and 3 are updated. This ends BFS from vertex 1 for level 0. Vertices 3 and 6 are processed concurrently at level 1 (available from Stack_Array[1]) in a similar

---

Algorithm 3.13: Parallel Betweenness Centrality Computation Algorithm

---

```
1  Parallel_BC(Graph(V,E)) {
2        // BC[v] is the betweenness centrality of vertex v.
3        // A number of initializations follow.
4        foreach( v In V ) in parallel {
5           |    BC[v] = 0;
6        }
7        foreach( v In V ){
8             // The above is a sequential loop. Therefore, separate prdecessor
9             // sets and stacks are not needed for each vertex. Parallel
10            // invocation of BFS for each vertex v will need too much memory.
11            foreach( w In V ) in parallel {
12               |    predecessor[w] = φ; σ[w] = 0;
13               |    distance[w] = −1;
14            }
15            σ[v] = 1; distance[v] = 0; level = 0; count = 1; Stack_Array[level] = φ;
                 Push(v,Stack_Array[level]); // Each entry of Stack_Array is a stack,
16            // and corresponds to vertices at a particular "level" of BFS.
17            // End of initializations. Level is BFS level.
18            // Begin parallel BFS of graph, compute shortest distance,
19            // predecessors, and no. of shortest paths (σ[]).
20            while( count > 0 ){
21                 count = 0;
22                 foreach( w In Stack_Array[level] ) in parallel {
23                      foreach( n In w.nbrs ) in parallel {
24                           atomic
25                           if( distance[n] == -1 ){
26                                count++;
27                                Push(n,Stack_Array[level+1]);
28                                distance[n] = level + 1;
29                           }
30                           end
31                           atomic
32                           if( distance[n] == level+1 ){
33                                σ[n] = σ[n] + σ[w];
34                                predecessor[n] = predecessor[n] ∪ {w};
35                           }
36                           end
37                      }
38                 }
39                 level = level + 1;
40            }
41            level = level - 1; // Now accumulate partial pairwise dependencies and add to BC.
42            foreach( s In V ) in parallel {
43               |    δ[s] = 0;
44            }
45            while( level > 0 ){
46                 foreach( u In Stack_Array[level] ) in parallel {
47                      foreach( t In predecessor[u] ){
48                           atomic
49                           |    δ[t] = δ[t] + σ[t]/σ[u] (1 + δ[u]);
50                           end
51                      }
52                      if( u ≠ v ){
53                         |    BC[u] = BC[u] + δ[u];
54                      }
55                 }
56                 level = level - 1;
57            }
58       }
59  }
```

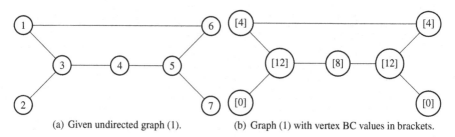

(a) Given undirected graph (1).   (b) Graph (1) with vertex BC values in brackets.

Fig. 3.6 Example for parallel computation of betweenness centrality

manner, with vertices 4, 2, and 5 being pushed into Stack_Array[2], to be processed at level 2 of BFS. All these operations are summarized in Tables 3.5, 3.6, and 3.7.

3.5.5 Parallel Union-Find and Applications

Sequential algorithms for disjoint set union and find operations have been well known for several years [64, 65]. They are used routinely in applications such as Kruskal's algorithm for computing minimum spanning trees. However, parallel versions of union and find operations are not as efficient as their sequential ones. Simple approaches to making these sequential algorithms into parallel ones by using locks on all update operations limits performance. Frameworks such as Galois [66] provide efficient implementations of concurrent Union-Find algorithms which use fine grained locks. Concurrent wait-free versions of these operations are reported in [67] and more recently in [68]. These are randomized concurrent algorithms and their performance is estimated to be very good with certain assumptions on randomization. Even though their performance may be reasonable in practice, sufficient experimentation with these algorithms has still not been carried out to verify the claims of theory. The following sections describe versions of wait-free Union-Find algorithms as described in [68]. They may be replaced by available implementations such as the ones in Galois [66] with almost no change in the structure of applications using them.

3.5.5.1   Concurrent Disjoint Set Union and Find Algorithms

The collection of sets that are dealt with in these algorithms are assumed to be disjoint, for example, collection of disjoint sets of vertices used in Kruskal's and Boruvka's MST algorithms. Each set in the collection is maintained as a tree with the root being the representative of the set ("ID" of the set). Parent pointers are used to implement the tree, with parent of the root pointing to itself. The MakeSet operation creates a collection of singleton sets. A find operation on a leaf of a tree

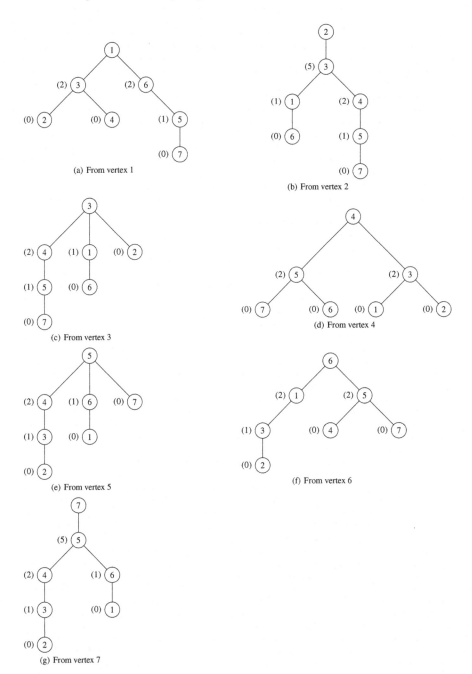

Fig. 3.7 Parallel BC computation—BFS level-by-level traversal from each node. Partial δ values are shown beside the nodes in parentheses

Table 3.5 Summary of *pred, dist*, and σ computations for vertex 1 in Fig. 3.6a

Vertex processed	BFS level	Operations	Stack_Array (SA) contents	Predecessor (Pred) set	Distance (dist)	σ
None		Push 1 onto SA	SA[level=0]=1			
1	0	Push 6,3 onto SA	SA[level=1]=6,3	Pred[6]=1	dist[6]=1	$\sigma[6]=1$
		Push 1 onto Pred		Pred[3]=1	dist[3]=1	$\sigma[3]=1$
3	1	Push 4,2 onto SA	SA[level=2]=4,2	Pred[4]=3	dist[4]=1	$\sigma[4]=1$
		Push 3 onto Pred		Pred[2]=3	dist[2]=2	$\sigma[2]=1$
6	1	Push 5 onto SA	SA[level=2]=4,2,5	Pred[5]=6	dist[5]=2	$\sigma[5]=1$
4	2	Nothing to do since neighbours 3,5 are already in SA				
2	2	Nothing to do since neighbour 3 is already in SA				
5	2	Push 7 onto SA	SA[level=3]=7	Pred[7]=5	dist[7]=3	$\sigma[7]=1$
7	3	Nothing to do since neighbour 5 is already in SA				

Table 3.6 δ-computations related to vertex 1 in parallel BC Algorithm

Vertex processed	BFS level	Pred	$\delta[]$
7	3	Pred[7]=5	$\delta[5] = \delta[5] + 1 + \delta[7] = 1$
5	2	Pred[5]=6	$\delta[6] = \delta[6] + 1 + \delta[5] = 2$
2	2	Pred[2]=3	$\delta[3] = \delta[3] + 1 + \delta[2] = 1$
4	2	Pred[4]=3	$\delta[3] = \delta[3] + 1 + \delta[4] = 2$
6	1	Pred[6]=1	$\delta[1]$ is not calculated
3	1	Pred[3]=1	As it is the source of BFS

traverses parent pointers till the root and returns the ID of the root. A union operation on two sets A and B combines the two trees corresponding to the two sets, by making the root of tree with a "lesser" ID for its root as the child of the tree with "greater ID" for its root. This assumes that the IDs of the elements have already been assigned to the elements and that they will not be changed any time. Elements may be assigned their IDs either as a total order or as a random permutation of a total order, from the range $1, \ldots, N$, with N being the total number of elements (such as number of vertices in a graph).

Table 3.7 δ-computations related to other vertices in parallel BC Algorithm

Vertex processed	$\delta[]$
1	$\delta[7] = 0, \delta[2] = 0, \delta[4] = 0, \delta[5] = 1, \delta[3] = 2, \delta[6] = 2$
2	$\delta[7] = 0, \delta[6] = 0, \delta[5] = 1, \delta[1] = 1, \delta[4] = 2, \delta[3] = 5$
3	$\delta[7] = 0, \delta[6] = 0, \delta[5] = 1, \delta[1] = 1, \delta[4] = 2, \delta[2] = 0$
4	$\delta[7] = 0, \delta[6] = 0, \delta[1] = 0, \delta[2] = 0, \delta[5] = 2, \delta[3] = 2$
5	$\delta[2] = 0, \delta[3] = 1, \delta[1] = 0, \delta[4] = 2, \delta[6] = 1, \delta[7] = 0$
6	$\delta[2] = 0, \delta[3] = 1, \delta[4] = 0, \delta[7] = 0, \delta[1] = 2, \delta[5] = 2$
7	$\delta[2] = 0, \delta[1] = 0, \delta[3] = 1, \delta[6] = 1, \delta[4] = 2, \delta[5] = 5$

Algorithm 3.14: `MakeSet()` Operation on a Single Element

```
1  MakeSet(Elem t) {
2      t.parent = t;
3      t.ID = AssignID();
4      // AssignID() may assign IDs either according to
5      // a total order or a random order.
6  }
```

Algorithm 3.15: `Find()` with Path Splitting

```
1  Find(Elem t) {
2      // Splitting replaces the parent of every node on
3      // the path by its grandparent.
4      Elem u = t, v, w;
5      while( True ){
6          // The loop exits when root is reached
7          v = u.parent;
8          w = v.parent;// w is grandparent of u.
9          if( v == w ){
10             return v;
11         }
12         else{
13             CAS(u.parent, v, w); u = v;
14         }
15     }
16 }
```

3.5.5.2 Finding Connected Components in Parallel Using Union-Find

Finding connected components using Union-Find is almost trivial. Start with each vertex as a component, consider each edge in parallel, if the vertices of the edge belong to different components, merge the two components into one. This sequence is repeated until no more Union operations are possible. Algorithm 3.18 describes this procedure.

Algorithm 3.16: Algorithm to Determine Two Elements Are
in the Same Set

```
1  SameSet(Elem t1, Elem t2) {
2      // Determines if t1 and t2 are in the same set.
3      // Returns TRUE if so, FALSE, otherwise.
4      Elem u = t1, v = t2;
5      while( True ){
6          // The loop exits when root is reached
7          u = Find(u);
8          v = Find(v);
9          if( u == v ){
10             return True;
11         }
12         // If u and v are not the same, then their sets might
13         // have been merged by another thread after the
14         // Find() operation. Checking if u.ID < v.ID and u is
15         // still a root is enough; return false if so.
16         // Even if root of v has changed, it could not have
17         // merged into tree of u since u.ID < v.ID. Iterate otherwise.
18         if( u.ID < v.ID && u == u.parent ){
19             return False;
20         }
21     }
22  }
```

3.5.5.3 Parallel MST Computation with Boruvka's Algorithm

Sequential version of Boruvka's algorithm for computing minimum spanning trees
(MST) has been described in Algorithm 2.8. Surprisingly, the algorithm is also
amenable for easy parallelization. A version using parallel Union-Find as the basis
is presented in Algorithm 3.19. The algorithm begins by creating a collection of
singleton sets out of the V, the set of vertices (Line 6). Each vertex x will have extra
properties, viz., $x.minweight$, $x.mintarget$, $x.minsource$, and $x.active$. $x.minweight$
stores the weight of the minimum weight edge going out of vertex x. $x.minsource$
and $x.mintarget$ store the source and target vertices of the minimum weight edge out
of vertex x. $x.active$ indicates whether or not a vertex is going to participate in the
set merge process (zero initially).

Lines 10–23 compute the minimum weight edge out of each vertex in a
component, and store the information in the appropriate properties of the root of the
component. Lines 25–33 choose the root of a component as the *active* root (*active*
= *1*)to initiate merger of the two components corresponding to the two vertices at
the end of an edge. The other end of the edge is made the passive root (*active* = *2*).
Lines 35–42 merge the end components of an edge and add the edge to the MST.
The algorithm loops until all the vertices are merged into a single component as
indicated by the absence of an edge between two components to be added to the
MST (Line 14).

Algorithm 3.17: Union Operation Using Index

```
1  Union( Elem t1, Elem t2 ) {
2        // Union returns true if t1 and t2 are in different sets
3        // and union is indeed performed. Otherwise, if t1 and t2 are in the
4        // same set, it returns false.
5        Elem u = t1, v = t2, t;
6        while( True ){
7            u = Find(u); v = Find(v);
8            if( u.ID < v.ID ){
9                if( CAS(u.parent, u, v) ){
10                   // If CAS is successful, union has been performed.
11                   // Set u (smaller ID) has been linked to set v (larger ID).
12                   // Otherwise, try again by going back in the loop.
13                   return True;
14               }
15           }
16           else{
17               if( u == v ){
18                   return False; // Both u and v are in the same set.
19               }
20               else{
21                   if( CAS(v.parent, v, u) ){
22                       // If CAS is successful, union has been performed.
23                       // Set v (smaller ID) has been linked to set u (larger ID).
24                       // Otherwise, try again by going back in the loop.
25                       return True;
26                   }
27               }
28           }
29       }
30 }
```

3.6   Graph Analytics on Distributed Systems

A computer cluster or distributed computer system consists of a cluster of inter-connected machines. Each machine has its own private memorymay have a GPU as an accelerator in addition to the CPU. So the cluster can be divided in to three categories:

- Distributed systems with CPUs
- Distributed systems with CPUs and GPUs.
- A single machine with multiple GPUs (multi-GPU machine).

Note that because GPUs often have separate memories than CPUs, a single machine with multiple GPUs leads to a distributed setup. Distributed graph processing requires graph to be partitioned into multiple subgraphs. Basic graph partitioning methods like *vertex-cut* and *edge-cut* were discussed in Sect. 1.4. More partitioning strategies exist. The *p-way* vertex-cut partitions edges equally to p

---

Algorithm 3.18: Finding Connected Components Using Parallel Union-Find

---

```
1  Parallel_Connected_Components(Graph G(V, E)) {
2         // Go on merging components of each vertex of an edge,
3         // until no more merges are possible.
4         foreach( v In V ) in parallel {
5             │  MakeSet(v);
6         }
7         while( True ){
8             changed = False;
9             foreach( Edge e:<v,w> In E ) in parallel {
10                │  if( Union(v,w) ){
11                │      // Iterate once more, even if one Union succeeds.
12                │      // Exit the loop only when all Unions fail.
13                │      // That happens only when all the vertices in a
14                │      // connected component have been merged into
15                │      // a single component.
16                │      changed = true;
17                │  }
18            }
19            if(changed == False) break;
20        }
21  }
```

---

machines. The same vertex exists in multiple subgraphs during a partitioning ((called *replication*). The replication factor is much more for *vertex-cut* compared to *edge-cut* partitioning. Another form of vertex cut called *constrained* vertex-cut is also available. This strategy requires the number of subgraphs to be close to a perfect square to achieve better partitioning. Hybrid graph partitioning methods which combine both vertex-cut and edge-cut are also used in the past [69, 70]. In a *cartesian* vertex-cut, the *master node* of the vertex has both the incoming and the outgoing edges. The remote vertices have incoming only or outgoing only edges. The *incoming edge-cut* imposes more restrictions. The incoming edges are assigned only to the master node while outgoing edges of a vertex are partitioned among different machines. The *outgoing* edge-cut is the mirror of the *incoming* edge-cut.

The distributed graph analytics requires communication between computing nodes. This is done with libraries such as the Message Passing Interface (MPI). The MPI library is discussed in detail in Sect. 1.6.5. The subgraphs are processed on different computing nodes. In a distributed execution, the mutable graph properties can be updated only using *commutative and associative* operations [71]. Programming distributed graph analytics on large graphs is challenging due to graph partitioning, minimized communication etc.

Algorithm 3.19: Parallel Version of Boruvka's Algorithm for Computing Minimum Spanning Tree

---

```
1  Parallel_Boruvka_MST(Graph G(V, E)) {
2      // Each component x will have x.minweight, x.mintarget, x.minsource,
3      // and x.active as properties.
4      foreach( v In V ) in parallel {
5          |  MakeSet(v);
6      }
7      changed = True;
8      while( changed ){
9          changed = False;
10         foreach( v In V ) in parallel {
11             p = Find(v); p.minweight = 0; p.active = 0;
12             p.mintarget = NULL; p.minsource = NULL;
13             foreach( w In v.neighbours ) in parallel {
14                 if( ! SameSet(v, w) ){
15                     change = True;
16                     atomic
17                         if( p.minweight > getweight(v, w) ){
18                             p.minweight = getweight(v, w);
19                             p.minsource = v; p.mintarget = w;
20                         }
21                     end
22                 }
23             }
24         }
25         foreach( v In V ) in parallel {
26             p = Find(v); q = Find(p.mintarget);
27             atomic
28                 if( !SameSet(p, q) ){
29                     if( (p.active == 0) && (q.active ≠ 1) ){
30                         p.active = 1; q.active = 2;
31                     }
32                 }
33             end
34         }
35         foreach( v In V ) in parallel {
36             atomic
37                 p = Find(v); q = Find(p.mintarget);
38                 if( p.active == 1 && q.active == 2 ){
39                     MST = MST ∪ {<p.minsource,p.mintarget>};
40                     Union(p,q); t = Find(v); t.active = 2;
41                 }
42             end
43         }
44     }
45 }
```

3.6.1 Distributed System with CPUs

The *Gather-Apply-Scatter* model, Vertex API based BSP model, and worklist model for distributed execution were discussed in Sect. 1.9.1. These models are implemented for CPU clusters. Minimizing the communication between subgraphs, maintaining proper work-balance etc. are required in a distributed computation. These are hard problems and *heuristics* are used to achieve suitable approximations in practice. The *MapReduce* framework is used for a rich set of large scale computation. The execution model of this framework provides inferior performance for graph analytics. The model is found suitable for applications such as SQL query processing.

In a BSP model of execution a vertex may be *active* or *inactive*. An *active* vertex performs computation, communication, and processes the received data. If there is mismatch in the amount of computation and communication among subgraphs, load-imbalance may occur. The load balancing in graph analytics is improved with dynamic strategies. Such a scheme does not rely fully on the input graph structure. It looks at how the computation happens at the runtime. Based on the runtime behaviour, graph partitions are modified by moving vertices and associated properties from one subgraph to another. Frameworks have considered sizes of the incoming and the outgoing messages, and time of each superstep in taking decisions at runtime [72].

There exists work which improve on top of the *Gather-Apply-Scatter* model of execution. In one extension a hybrid vertex-cut and edge-cut partitioning with separate communication engine is used. Another extension uses agent-graph model to partition graphs, and uses scatter-agent and combine-agent to reduce communication overhead [73]. Domain Specific Languages (DSLs) for graph analytics exist on CPUs. They different in the data structures and language constructs [74–76]. The DSLs which supports incremental dynamic algorithm have also been implemented [26, 27]. DSL with different way of programming where *algorithm* (computation) and *schedule* (how to compute) are specified separately are proposed recently. Separate languages are provided for *algorithm* and *schedule* [76].

Fault tolerance is an important requirement in graph analytics. Frameworks which reduce overhead on fault tolerance make the computation to happen from scratch [77]. There also exist frameworks which exploit check-pointing resulting in only a small performance overhead [78]. and frameworks that perform replication based fault-tolerance [79] Recent frameworks running on a CPU cluster, in which computation restarts only from a *safe point*, guarantee correct output [80].

3.6.2 Distributed System with CPUs and GPUs

Graph analytics on cluster with each node having a CPU and a GPU have been explored. It need to be noted that there is major difference in architecture style of the two types of devices. So this becomes a heterogeneous distributed system. A salient feature of a heterogeneous system is that the computation amenable to CPU-based asynchronous processing can be executed on CPUs, while the data-parallel processing can be executed on GPUs, achieving the best of both the worlds. Unfortunately, the communication overhead may becomes a bottleneck in heterogeneous systems. There exist frameworks which optimize communication by using *broadcast* message from *master* node to *remote* nodes, and *reduce* message for communication from *remote* nodes to *master* node [81].

There exist DSLs which support CPU and GPU clusters [27]. Such a DSL reports huge communication overhead for computation on GPU clusters as communication needs to happen from GPU to GPU through CPU memory. But computation on GPUs finishes relatively faster than the inter-device communication. In effect, communication time is less for CPU and computation time is less for GPU.

3.6.3 Multi-GPU Machine

The frameworks which supports multi-GPU machines have been developed in the past [82]. An advantage of using multi-GPU machine is its power-efficiency as well as cost-per-watt rating. Multi-GPU machines allow the computation to be largely partitioned across GPUs, with CPU only coordinating across devices. The DSL which supports CPU and GPU cluster also supports multi-GPU machines [26].

Chapter 4
Graph Analytics Frameworks

Frameworks take away the drudgery of routine tasks in programming graph analytic applications. This chapter describes in some detail, the different models of execution that are used in graph analytics, such as BSP, Map-Reduce, asynchronous execution, GAS, Inspector-Executor, and Advance-Filter-Compute. It also provides a glimpse of different existing frameworks on multi-core CPUs, GPUs, and distributed systems.

4.1 Introduction

The importance of graph algorithms for different applications was discussed in the previous chapters. Programming these algorithms in an efficient manner on heterogeneous hardware is challenging. Novel algorithms targeting multi-core machines have been proposed. Typical examples among elementary algorithms being Δ-stepping SSSP [10], divide and conquer SCC [34, 60] and Boruvka's MST computation for parallel machines. Graph algorithms are efficient on GPUs and GPUs follow the SIMT architecture. However, rewriting a graph algorithm to suit a particular architecture is not only challenging but also impractical. One way to tame this challenge is to provide high level abstractions for graph algorithms that would be automatically translated to efficient backend code. This way, the complexity of programming graph analytics applications is reduced substantially with abstractions. The abstraction can hide the hardware architecture, provide efficient (sequential or parallel) implementation of data structures (e.g.,Graph, Worklist, Set), graph partitioning heuristics, and support parallel constructs. Availability of such high-level abstractions allows the programmer to focus more on the conversion of an algorithm to a program in a programming language of choice.

Such a requirement gave rise to graph analytics frameworks. They provide high level abstractions in a high level language (e.g., C++, Java). Literature is aplenty

© Springer Nature Switzerland AG 2020
U. Cheramangalath et al., *Distributed Graph Analytics*,
https://doi.org/10.1007/978-3-030-41886-1_4

with several graph analytics frameworks. While all the graph analytics frameworks
provide such a functionality at the macroscopic level, they differ in multiple aspects.
Typically, a framework supports either synchronous execution [14], or asynchronous
execution [17], or both [11]. This respectively means, the workers (threads or
processes) execute parallel steps and coordinate to wait for others to finish each
step, or run completely independently of other threads, or use a combination of
the two. They also differ in the programming styles that they support. *Gather-
Apply-Scatter (GAS)* [11], *inspector-executor* [66], *advance-filter-compute* [12] and
Map-Reduce [83] are among the well known frameworks that have been proposed
in the past for graph analytics. Each style dictates a manner in which the underlying
computation must be encoded. The framework enforces explicit control-flow for
programming algorithms. Nuances of a particular computing system are handled
internally by the framework (e.g., graph partitioning and communication in a
distributed setup).

Graph algorithm implementation on GPUs started with handwritten codes. Efficient implementations of algorithms such as Breadth First Search (BFS) and Single
Source Shortest Path (SSSP) on GPUs were proposed several years ago [84, 85].
The BFS implementation from Merrill et al. [57] is novel and efficient, which
exploits worklist-based processing to improve work-efficiency of the implementation. Efficient implementations of other algorithms on GPUs such as n-body
simulation [86], betweenness centrality [87], data flow analysis [88] and control-
flow analysis [89] were also proposed later. Efficient graph analytics frameworks
targeting GPUs [12, 90] and multi-GPU machines [20, 91] are available now.

4.2 Frameworks—Merits and Demerits

Programming a parallel graph algorithm even in a high level language is challenging
and error prone. Several of these algorithms rely on efficient data structures.
Examples are *Union-Find* data structure for Boruvka's MST, *Buckets* for Δ-
Stepping SSSP. So a framework or a library with efficient implementation of the
required data structures and parallel constructs makes the task manageable for the
programmer. In other words, programming using a framework or a library is less
error prone and more efficient.

Efficient graph analytics libraries [92–94] are often used by frameworks as their
back-ends. Typical example include Galois [66] which uses the Boost [94] library. If
a library is used for programming graph algorithms, the programmer must define the
flow of computation. In addition, the programmer is required to decide *where* to call
a library function. Graph analytics libraries provide some abstractions as part of their
API using other generic libraries such as MPI [5] and pthreads. However, a graph
algorithm's implementation is restricted by the vocabulary (functions) provided by
the library. Optimizations across functions often becomes difficult.

In contrast, frameworks make programming an algorithm much easier compared
to that done using a library. Frameworks provide a generic API [11, 12] for the

implementation of the algorithm. Programmer is asked to implement different parts of the algorithm (computation, communication, aggregation) in specific functions of the API. So the flow of computation is restricted by the framework, thereby providing modularity and better programming style, and making it comfortable to program algorithms. Frameworks internally manage computing threads, communication etc. They provide efficient implementation of data structures required for the algorithm implementation. The preferred abstraction and requirements for multiple algorithms may be available in a framework. Preferences may include synchronous or asynchronous execution, single multi-core CPU or GPU, or distributed computation, fault tolerance, heterogeneous (CPU and GPU) computing, etc.

A framework typically does not provide the features and facilities of a general purpose programming language and its compiler. This includes static program analysis and type checking. A domain specific language (DSL) provides better abstraction and programming style than a framework. The availability of higher level abstractions, typechecking, program analysis, and better optimizations makes DSLs very attractive. The layers of abstractions follow a hierarchy of DSL, framework, library, and generic library. For example, Boost uses MPI, Galois uses Boost, and Falcon [26] DSL uses Galois for Δ-Stepping implementation on multi-core CPUs.

4.3 Models for Graph Analytics

In this section we discuss various popular execution models for graph algorithms and the frameworks and DSLs that provide them.

4.3.1 Bulk Synchronous Parallel (BSP)

A brief description of BSP was provided in 1.9.1.1. The BSP model performs computations in supersteps [13]. The reader may recall that the each superstep consists of three parts: Computation, Communication and Barrier Synchronization. A brief decsription of important frameworks that follow BSP are provided below.

4.3.1.1 Pregel

Pregel [14] is a proprietary graph analytics framework from Google, that follows the BSP model. The input graph G(V, E) can have a set of mutable properties associated with vertices and edges. In each *superstep*, vertices carry out the computation in parallel. The algorithmic logic must be programmed in the API functions of Pregel (see Algorithm 4.1). A vertex modifies the mutable properties of its neighbouring vertices and edges, sends messages to vertices, receives messages from vertices, and if required changes the topology of the graph. A typical example for a mutable property is the vertex property *dist* in the SSSP computation. All active vertices perform computation and all the vertices are as set active initially.

A vertex deactivates itself by calling the *VoteToHalt()* API function and it gets reactivated automatically when a message is received from another vertex. Once all the vertices call the *VoteToHalt()* function and no message is sent across vertices, the algorithm terminates. Algorithmic logic of the computation is required to be coded in the *Compute()* function. The message, vertex and edge data types are specified as templates. Different template types may be chosen for different algorithms. A programmer is required to override the *Compute()* function, which will be run on all the active vertices in each superstep, and code the algorithmic logic of the computation in it.

Algorithm 4.1: Vertex API of Pregel

```
1  template <typename VertexValue, typename EdgeValue, typename
   MessageValue>
2  class Vertex {
3  public:
4      virtual void Compute(MessageIterator* msgs) = 0;
5      const string& vertex_id() const;
6      int64 superstep() const;
7      const VertexValue& GetValue();
8      VertexValue* MutableValue();
9      OutEdgeIterator GetOutEdgeIterator();
10     void SendMessageTo(const string& dest_vertex, const MessageValue&
   message);
11     void VoteToHalt();
12 }
```

4.3.1.2   GPS

The GPS [95] framework follows the BSP model of execution and incorporates incremental modifications to Pregel. GPS repartitions the vertices of the graph across the nodes of a distributed system during the computation, based on message-sending patterns. It tries to combine two vertices on two different nodes which send messages among themselves frequently during computation. The GPS framework provides the *Master.Compute()* function in addition to the *Vertex.Compute()* function of Pregel. The *Master.Compute()* function is called at the beginning of each superstep. The Master class has access to merged global objects and *Compute()* function can update global objects.

4.3.1.3   Pregelix

Pregelix [96] uses set-oriented, iterative dataflow approach for the implementation of the Pregel programming model. Pregelix combines the messages and vertex states in a computation into a tuple. Query evaluation technique is then used to execute the user program. Pregelix models the Pregel semantics as a query plan and implements the semantics as an iterative dataflow of relational operators. Message exchange is taken as the join followed by a group-by operation. Pregelix offers a set of alternative physical evaluation strategies for different workloads.

4.3.1.4   Ligra

Ligra [97] is a graph processing framework for shared-memory multicore machines. The framework comes with an API that has two routines, one each for mapping edges and vertices respectively. The routines can be applied to any subset of vertices. This is useful for graph traversal algorithms that operate on subsets of vertices. Ligra+ [98] focuses on compression methods to reduce space complexity for in-memory graph algorithms. It is able to reduce the storage requirement by 50% for storing a graph in volatile memory. Ligra+ is slightly faster than Ligra when experimented on a machine with 40 cores (with hyperthreading).

4.3.1.5   Apache Giraph and Apache Hama

Apache Giraph [99] is an open source framework based loosely on the Pregel model. It also supports programming in the *Map-Reduce* model. Apache Giraph is extended to handle large scale graphs in   [15]. The framework was evaluated for graphs with more than 6 Billion edges. Apache Hama [100] is a distributed computing framework based on the BSP model. It follows a Pregel-like computation model for graph analytics. The SSSP algorithm in Hama is shown in Algorithm 4.2.

Algorithm 4.2: SSSP Algorithm in Apache Hama

```
1           compute(messages)
2  if getSuperstepCount() == 0 then
3      setValue(INFINITY );
4  end if
5  int minDist = isStartVertex() ? 0 : INFINITY;
6  for int msg : messages do
7      minDist = min(minDist, msg);
8  end for
9  if minDist < getValue() then
10     setValue(minDist);
11     for Edge e: getEdges() do
12         sendMessage(e, minDist + e.getValue())
13     end for
14 end if
15 voteToHalt();
16          combine(messages)
17 return min(messages);
```

Algorithm 4.2 works on a graph G(V, E). The computation happens on *messages* in the *compute()* function. In the first superstep, the *value* (distance) of each vertex is set to *INFINITY* (see Lines 2–4). The variable *minDist* is local for each *vertex*. The value of *minDist* is set to zero for *StartVertex* and *INFINITY* for all other vertices in each iteration (see Line 5). The *getValue()* function returns the *value* of the vertex. In the following supersteps each vertex attains the smallest *value* of the messages received. If the received message *value* for a vertex is less than its current *value*, the vertex updates its *value* using the *setValue()* function. This reduces the vertex *value*. If the value gets reduced for a vertex *u*, then a message with a *value u.value* + *e.weight* is sent to all the vertices *v*, where $u \rightarrow v \in E$, using the *SendMessage()* function. At the end of each iteration, the vertex makes itself inactive by calling *voteToHalt()* function. If a vertex receives a message, it gets reactivated for the next superstep. The number of messages sent over the network is reduced using the function *combine()*, which takes the minimum *value* of messages for each destination vertex and sends only one message with the minimum *value*. Computation stops when no *message* is sent in a superstep denoting a fixed point.

4.3.1.6 Totem

The BSP execution model is used in GPUs and multi-GPU machines also. Totem [20] is a framework for graph processing on a heterogeneous and multi-gpu system. It follows the BSP execution model. It supports running algorithms in a distributed fashion among multi-core CPU and multiple GPUs of the multi-GPU machine. The graph object is partitioned and each subgraph is stored on the devices used for computation. Totem stores graphs in the Compressed Sparse Row (CSR) format. It partitions graphs in a way similar to edge-cut partitioning.

Totem uses the buffers *outbox* and *inbox* on each device for communication. The *outbox* buffer is allocated with space for each remote vertex, while the *inbox* buffer has an entry for each local vertex that is a remote vertex in another subgraph on a different device. Totem partitions a graph onto multiple devices, with less storage overhead. It aggregates boundary edges (edges whose vertices belong to different master devices) to reduce communication overhead. It sorts the vertex ids in the *inbox* buffer to have better cache locality. Totem has inbuilt benchmarks which the user can specify as a numerical value. A user can also specify how many GPUs to use, the percentage of the graph that should go to GPU etc., as command line arguments. Such heterogeneous computing is useful as some algorithms do not perform well on GPUs alone. Examples are SSSP, BFS etc., on GPU for road networks. This happens as road networks have a large *diameter* and hence less parallelism is possible. A user can dictate that an algorithm should be executed either on CPU or on GPU, based on the type of input and/or algorithm.

The basic structure of the Totem framework is shown in Algorithm 4.3, where a Totem benchmark defines the parameters in the *totem_config* class. A new benchmark should use this template and place the algorithmic logic properly in the functions in the `struct totem_config`. The totem execution engine will take

---

Algorithm 4.3: Totem Engine Structure

```
1  struct totem_config config = {
2      graph,
3      partitioning_algo(),
4      init_func(),
5      kernel_func(),
6      msg_reduce_func(),
7      finalize_func(),
8  };
9  totem_config(&config);
10 totem_execute();
```

---

care of the runtime environment which involves computations and communications. The benchmark is executed with a call to *totem_config*, followed by *totem_execute()*. Totem does not have a feature to run multiple algorithms on the same input graph using different devices on a machine. Such a feature is useful when multiple properties of a single graph object are required to be computed by running different algorithms (e.g. SSSP, BFS, Pagerank etc.) on the same input graph using multiple devices (GPUs and single CPU).

There are several other graph analytics frameworks which follow the BSP execution model [101, 102].

4.3.2 Map-Reduce

Hadoop [103] supports Map-Reduce processing of graphs and uses the Hadoop distributed file system (HDFS) for storing data. HaLoop [83] is a framework which follows the Map-Reduce pattern with support for iterative computation and with better caching and scheduling methods. Twister [104] is also a framework which follows the Map-Reduce model of execution. Pregel-like systems can outperform Map-Reduce systems in graph analytic applications.

4.3.3 Asynchronous Execution

GraphLab [17] is an asynchronous distributed computation framework in which vertex programs have access to a distributed graph with data stored at every vertex and every edge. Each vertex program may directly access information on the current vertex, adjacent edges, and adjacent vertices irrespective of the edge direction. Vertex programs can schedule neighboring vertex programs to be executed in the future. GraphLab ensures serializability[1] [105, 106] by preventing

---

[1]Serializability ensures that executing concurrent transactions with a particular schedule is equivalent to a schedule that executes the transactions serially in some order.

neighboring program instances from running simultaneously. In a distributed setup, it is crucial to minimize inter-node communication, which often happens via sending and receiving messages. By eliminating messages, GraphLab isolates user-defined algorithms from the movement of data, thereby allowing the system to choose when and how to move the program state. GraphLab uses *edge-cut* partitioning of graphs and for a vertex *v*, all its outgoing edges will be stored in the same node.

Algorithm 4.4: GraphLab Execution Model

---

1 **Input:** Data Graph G = (V, E, D) // D is the data table
2 **Input:** Initial vertex worklist T = $\{v_1, v_2,...\}$
3 **Output:** Modified Data Graph G = (V, E, D')
4 **while**($T \neq \phi$){
5   v ← popNext(T) // *v* is removed from *T*.
6   (T', S_v) ← update(v, S_v)
7   T ← T ∪ T'
8 }

Algorithm 4.5: PageRank Calculation in GraphLab

---

1 **Input:** Vertex data $pr(v)$, Vertex Scope S_v, Edge data $weight(u, v) : u \in Nbrs(v)$
2 **Input:** Neighbour vertex data $pr(u) : u \in Nbrs(v)$, α, ξ
3 $pr_{old}(v)$ = pr(v) // Save old PageRank
4 pr(v) = α / n
5 **foreach**($u \in Nbrs(v)$){
6   pr(v) = pr(v) + (1 - α) * weight(u, v) * pr(u)
7   **if** ((*abs (pr(v)* - $pr_{old}(v)$))) > ξ){
8   // Schedule neighbors to be updated
9   return u: u ∈ Nbrs(v)
10   }
11 }

The execution model of GraphLab is shown in Algorithm 4.4. The *data graph* G(V, E, D)[2] (Line 3) of GraphLab stores the program state. A programmer can associate data with each vertex and edge based on the algorithm requirement. The *update* function (Line 6) of GraphLab takes as input, a vertex *v* and its *scope* S_v (data stored in *v*, its adjacent vertices and edges). It returns modified scope S_v and a set of vertices *T'* which require further processing. The set *T'* is added to the set *T* (Line 7), so that it can be processed in the upcoming iteration. Algorithm terminates when *T* becomes empty (Line 4).

ASPIRE [107], KLA [108], CoRAL [109] and Wang et al. [110] describe other examples of asynchronous frameworks.

---

[2]V is the vertex set, E is the edge set and D is the data table.

4.3.4   External Memory Frameworks

Very large graphs do not fit in RAM and hence, external memory algorithms partition the graphs into smaller chunks such that each chunk fits into memory and can be processed. The computed values are then used to process the next chunk, and so on. The GraphChi [58] framework processes large graphs using a single machine, with the graph being split into parts, called *shards*, loading shards one by one into RAM and then processing each shard. It uses Gauss–Seidel iterative computation based on a parallel sliding window. Such a framework is useful in the absence of distributed clusters. The parallel sliding window technique was experimented with elementary graph algorithms [111] also. The Turbo-Graph framework [112] proposes the *pin-and-slide* execution model for parallel graph analytics using external memory. It supports graph analytics on multi-core CPU with FlashSSD I/O devices. The I/O operations and computations are performed asynchronously to provide more efficiency. TurboGraph has been shown to outperform GraphChi. TurboGraph is extended for external memory algorithms on CPU clusters [113]. The extension supports multiple levels of parallel and overlapped processing for efficient usage of multi-core CPUs, hard-disks, and network.

4.3.5   Gather-Apply-Scatter Model

Several graph computations can be naturally modeled using the *Gather-Apply-Scatter* (GAS) model. The GAS model is composed of functions *gather, sum, apply* and *scatter* which are invoked by the runtime engine. The *gather* function is invoked on all the adjacent vertices of a vertex u. The *gather* function takes as arguments, the data on an adjacent vertex and an edge, and returns an accumulator specific to the algorithm. The result is combined using the commutative and associative *sum* operation (which could also be algorithm-specific). The final gathered result a_u is passed to the *apply* phase of the GAS model. The pseudo code for such a computation is presented in Algorithm 4.6. The *scatter* function is invoked in parallel on edges adjacent to a vertex u, producing new edge values $D_{(u,v)}$. The *nbrs* in the *scatter* and *gather* phases can be *none, innbrs, outnbrs,* or *allnbrs*.

Algorithm 4.7 shows SSSP computation in PowerGraph [114] framework which follows the GAS model. The gather phase combines inbound messages. The apply phase consumes the final message sum and updates the vertex property. The scatter phase defines the message computation for each edge. Algorithm 4.8 shows the implementation of *Pagerank* in the GAS model.

Algorithm 4.6: Gather-Apply-Scatter Vertex Program

```
1  interface GASVertexProgram(u) {
2      // Run on gather_nbrs(u)
3      gather($D_u$, $D(u, v)$, $D_v$ ) → Accum
4      sum(Accum left, Accum right) → Accum
5      apply($D_u$, Accum) → $D_u^{new}$
6      // Run on scatter_nbrs(u)
7      scatter($D_u^{new}$, D(u, v), $D_v$ ) → ($D_{(u,v)}^{new}$, Accum)
8  }
```

Algorithm 4.7: PowerGraph: SSSP Computation

```
1  Gather($D_u$, $D_{(u,v)}$, $D_v$ ):
2      return $D_v + D_{(v,u)}$
3  Sum(a, b):
4      return min(a, b)
5  Apply($D_u$, new_dist):
6      $D_u$ = new_dist
7  Scatter($D_u$, $D_{(u,v)}$, $D_v$ ):
8      if (changed($D_u$)) Activate(v)
9      if (increased($D_u$)) return NULL
10     else return $D_u + D_{(u,v)}$
```

Algorithm 4.8: Pagerank Algorithm in Gather-Apply-Scatter Model

---
```
1  Gather (a: Double, b: Double) {
2  |    return a + b
3  }
4  Apply(v, msgSum) {
5  |      PR(v) = 0.15 + 0.85 * msgSum
6  |      if (converged(PR(v))) voteToHalt(v)
7  }
8  Scatter(v, j) {
9  |    return PR(v) / NumLinks(v)
10 }
```

4.3.6 Inspector-Executor

The execution time of a topology-driven graph algorithm depends considerably on graph topology, but not primarily on vertex or edge properties of the graph. Static program analysis will not be able to optimize topology-driven algorithms. Runtime dependences between vertices and edges enable some optimizations [115]. The *inspector* performs preprocessing to reduce communication volume and the number of messages between processors. After the communication phase, each processor starts the computation (*executor*). The remote data received by each processor is kept in its receive buffer. The *executor* uses a local copy of the accumulated remote data, and the received copy of remote data in the current iteration. After the computation, the values are copied to the local copy of the accumulated remote data.

Galois [66] uses the *inspector-executor* execution model for topology-driven graph algorithms targeting multi-core CPUs. It supports mutation (morphing) of graph objects via *cautious* speculative execution.[3] Galois uses a data-centric formulation of algorithms called *operator formulation*. Galois defines:

- *Active Elements*: the vertices or edges where computation needs to be performed at a particular instance of program execution.
- *Neighborhood*: the vertices or edges which are read or written by active elements in a particular instance of execution.
- *Ordering* of the active elements present at a particular instance of program execution.

In *unordered* algorithms, active elements can be processed (e.g., Delaunay Mesh Refinement, Minimum Spanning Tree) in any order, whereas in *ordered* algorithms, elements are processed in a particular order (e.g, Δ-Stepping SSSP).

Galois uses a *worklist* based execution model, where all the active elements are stored in a worklist and they are processed either in ordered or unordered fashion using a `foreach` operator. During the processing of active elements, new active elements are created, and these will be processed in the following rounds of computation. Computation stops when all the active elements have been processed and no more active elements are being created, denoting a fixed point.

Algorithm 4.9 shows the pseudo-code for Δ-Stepping SSSP implementation in Galois. Galois uses an *order by integer metric (OBIM)* bucket for Δ-stepping implementation of SSSP as declared in Line 26. The operator in the *InitialProcess* structure reduces the distance of the neighbours of the source vertex and adds to the OBIM buckets (Lines 19–21). This function is called from SSSP class in Line 29. Then the parallel *for_each_local* iterator of Galois calls the operator of Process structure in Line 30. This calls the operator of Process defined in Lines 9–11. The parallel iterator completes once all buckets are free and the SSSP distance of all the vertices are computed. The *relaxNode()* and *relaxEdge()* functions are not shown in the algorithm. They are used to reduce the distance values of vertices as done in other SSSP algorithms.

Galois also supports mutation of graph objects using cautious morph implementations (Delaunay Mesh Refinement and Delaunay Triangulation) and also algorithms based on mesh networks. Galois does not support multiple graph objects. Programming a new benchmark in Galois requires much effort, as understanding the C++ library and parallel iterators is more difficult compared to a DSL based approach.

---

[3]Morph algorithms can be classified as cautious, if the algorithms read all the neighborhood elements before modifying any of them.

Algorithm 4.9: SSSP in Galois C++ Framework

```
1  struct UpdateRequest {
2       Vertex n;
3       Dist w;
4  };
5  typedef Galois::InsertBag<UpdateRequest>Bag;
6  struct Process {
7       AsyncAlgo* self;
8       Graph& graph;
9       void operator()(UpdateRequest& req, Galois::UserContext<UpdateRequest>& ctx) {
10          self->relaxNode(graph, req, ctx);
11      }
12 };
13 struct SSSP{
14     struct InitialProcess {
15         AsyncAlgo* self;
16         Graph& graph;
17          Bag& bag;
18         Node& sdata;
19          void operator()(typename Graph::edge_iterator ii) {
20              self->relaxEdge(graph, sdata, ii, bag);
21          }
22     };
23     void operator()(Graph& graph, GNode source) {
24         using namespace Galois::WorkList;
25         typedef dChunkedFIFO<64> Chunk;
26         typedef OrderedByIntegerMetric<UpdateRequestIndexer<UpdateRequest>,
   Chunk, 10> OBIM;
27         Bag initial;
28         graph.getData(source).dist = 0;
29         Galois::do_all( graph.out_edges(source, Galois::MethodFlag::NONE).begin(),
   graph.out_edges(source, Galois::MethodFlag::NONE).end(), InitialProcess(this, graph,
   initial, graph.getData(source))));
30         Galois::for_each_local(initial, Process(this, graph), Galois::wl<OBIM>());
31     }
32 };
```

4.3.7 Advance-Filter-Compute

This model defines *advance*, *filter*, and *compute* primitives which operate on *frontiers* in different ways. A *frontier* is a subset of edges or vertices of the graph which is actively involved in the computation. The Gunrock [116] framework follows this execution model. Gunrock provides data-centric abstraction for graph operations at a higher level which makes programming graph algorithms easy. Gunrock has a set of API to express a wide range of graph processing primitives and targets Nvidia GPUs. Each operation in this model can be of the following types:

- An *advance* operation creates a new frontier using the current frontier by visiting the neighbors of the current frontier. This operation can be used in algorithms such as SSSP and BFS which activate subsets of neighbouring vertices.
- The *filter* operation produces a new frontier using the current frontier, but the new frontier will be a subset of the current frontier. An algorithm which uses such a primitive is the Δ-Stepping SSSP.
- The *compute* operation processes all the elements in the current frontier using a programmer defined computation function and generates a new frontier.

The SSSP computation in Gunrock is shown in Algorithm 4.10. It starts with a call to *SET_PROBLEM_DATA()* (Lines 1–6) which initializes the distance *dist* to ∞ and predecessor *preds* to NULL for all the vertices. This is followed by *dist* of *root* node being made to 0. Then the *root* node is inserted into the worklist *frontier*. The computation happens in the while loop (Lines 20–24) with consecutive calls to the functions *ADVANCE* (Line 21), *FILTER* (Line 22) and *PRIORITYQUEUE* (Line 23). The *ADVANCE* function with the call to *UPDATEDIST* (Lines 7–10), reduces the distance of the destination vertex *d_id* of the edge *e_id* using the value *dist[s_id]+weight[e_id]* where *s_id* is the source vertex of the edge. All the updated vertices are added to the frontier for processing in the next iterations. Then the *ADVANCE* function calls *SETPRED* (Lines 11–14) which sets the predecessor in the shortest path of vertices from the root node. The *FILTER* function removes redundant vertices from the frontier using a call to *REMOVEREDUNDANT*. This reduces the size of the worklist frontier which will be processed in the next iteration of the while loop. Computation stops when *frontier.size* becomes zero.

Programs can be specified in Gunrock as a series of bulk-synchronous steps. Gunrock also looks at GPU specific optimizations such as kernel fusion. Gunrock provides load balance on irregular graphs where the *degree* of the vertices in the *frontier* can vary a lot. This variance is very high in graphs which follow the power-law distribution. Instead of assigning one thread to each vertex, Gurnock loads the neighbor list offsets into the shared memory, and then uses a Cooperative Thread Array (CTA) to process operations on the neighbor list edges. Gunrock also provides *vertex-cut* partitioning, so that neighbours of a vertex can be processed by multiple threads. Gunrock uses a priority queue based execution model for SSSP implementation. Gunrock was able to get good performance using the execution model and optimizations (mentioned above) on a single GPU device.

4.4 Frameworks for Single Machines

Modern CPUs are equipped with multiple cores with MIMD architecture. Each CPU has a large volatile memory and a barrier across all threads can be implemented quite easily, as it follows the MIMD model. In comparison, GPU devices have a massively parallel architecture and follow the SIMT execution model. A GPU device is attached as a separate computing unit on a CPU. For example, the

---

Algorithm 4.10: SSSP Algorithm in Gunrock

```
1  procedure SET_PROBLEM_DATA (G, P, root)
2        P.dist[1..G.verts] ← ∞
3        P.preds[1..G.verts] ← NULL
4        P.dist[root] ← 0
5        P.frontier.Insert(root)
6  end procedure
7  procedure UPDATELDIST (s_id, d_id, e_id, P )
8        new_dist ← P.dist[s_id] + P.weights[e_id]
9        return new_dist <atomicMin(P.dist[d_id], new_dist)
10 end procedure
11 procedure SETPRED (s_id, d_id, P )
12        P.preds[d_id] ← s_id
13        P.output_queue_ids[d_id] ← output_queue_id
14 end procedure
15 procedure REMOVEREDUNDANT (node_id, P )
16        return P.output queue_id[node_id] == output_queue_id
17 end procedure
18 procedure SSSP(G, P, root)
19        SET_PROBLEM_DATA (G, P, root)
20        while P.frontier.Size() >0 do
21              ADVANCE (G, P, UPDATEDIST, SETPRED)
22              FILTER (G, P, REMOVEREDUNDANT)
23              PRIORITYQUEUE (G, P )
24        end while
25 end procedure
```

Nvidia K-80 GPU has 4992 cores, 24 GB device memory and a base clock rate of 560 MHz. GPUs are also used for General Purpose computing (GPGPU), apart from their extensive usage in graphics platforms. Graph algorithms are irregular, require atomic operations, and can result in thread divergence when executed on a streaming multiprocessor (SM). Threads are scheduled in multiple thread blocks. Each thread block is assigned a streaming multi-processor (SM). The Nvidia K-80 GPU has 26 SMs, each with 192 streaming processors (SP) (thus, $26 \times 192 = 4992$ cores).

A barrier for threads within a thread block is possible in GPU as threads in a thread block are scheduled on the same SM. A global barrier across all threads in a CUDA kernel which spans multiple thread blocks is not available in a GPU directly. This needs to be implemented in software [117] and has higher overheads. This will force each thread in a thread block to process multiple elements using a `for` loop, so that the total number of threads is not huge. Before a computation, data needs to be copied from the non-volatile storage of a computer to the volatile memory of its CPU, and then to the volatile memory of the GPU. Writing an efficient GPU program requires a deep knowledge of the GPU architecture, so that the algorithm can be implemented with less thread divergence, fewer atomic operations, coalesced access etc. The performance issues of graph analytics on GPU are explored in [118].

4.4.1 Multi-Core CPU

Ringo [119] is a system for analysis of large graphs (with hundreds of millions of edges) on a multi-core CPU. Ringo has a tight integration between graph and table processing and efficient conversions between graphs and tables. It supports various types of graphs. Ringo runs on a multi-core CPU machine with a large main memory, and outperforms distributed systems on all input graphs.

4.4.2 Single GPU

4.4.2.1 IrGL

IrGL [120] is a framework for single GPUs. It implements three optimizations, namely, *iteration outlining*, *cooperative conversion* and parallel execution of nested loops. IrGL is an intermediate code representation, on which these optimizations are applied and CUDA code is generated from it. *Iteration outlining* moves the iterative loop from the *host* code to the *device* code and this eliminates the performance bottleneck associated with multiple kernel calls in an iterative loop. *Cooperative conversion* reduces the total number of atomic operations by aggregating functions over thread, warp and thread-block level.

IrGL follows the Bulk-Synchronous-Parallel model and provides the constructs, `ForAll`, `Iterate`, `Pipe`, `Invoke`, `Respawn` etc. It provides nested parallelism for its parallel `ForAll` construct and provides a *retry* worklist to the kernel which is hidden from the programmer. A `Pipe` statement organises execution of a pipeline of kernels, each of which feeds its output to the next one in the pipeline. The `Pipe` statement can be invoked with an optional argument *Once*, in which case, the kernels inside the `Pipe` statement block will be executed once. IrgL has a `Respawn` statement which adds an element to the *retry* worklist.

Algorithm 4.11 shows a Δ-stepping SSSP implementation in IRGL for a GPU [121]. The INIT function which is called using the `Invoke` statement in Line 12 initializes the SSSP computation by making the distance of all the vertices ∞. It then makes the distance of the source vertex zero, and adds it to the worklist using the *push* operation. The INIT function will be called only once as it is enclosed inside the `Pipe` *Once*. The SSSP function (Lines 1–10) is called using the `Invoke` statement in Line 14). SSSP function call is nested inside `Pipe` (without the keyword `Once`) and SSSP has a `Respawn` statement in Line 4, which adds elements to the *retry* worklist. If the distance value is greater than the current delta value, it is added to the worklist (different from the *retry* worklist) using the *push* operation (Line 6). The `Respawn` operation makes the SSSP call loop until the *retry* worklist (bucket with current Δ value) becomes empty. Duplicate elements are then removed from other buckets and Δ is incremented. The `Pipe` loop (Lines 13–17) exits when the worklist (all buckets) becomes empty.

IrGL provides worklist based implementation of algorithms where the *retry* worklist is transparent to the programmer and constructs like `Respawn`, `Pipe` and `Iterate` are used to process the elements. The `ForAll` statement iterates over all the elements of an object in parallel (which is given as its argument). IrGL does not provide any support for clusters of GPUs.

Algorithm 4.11: SSSP Using `Pipe` Construct in IrGL

```
 1  Kernel SSSP(graph, delta) {
 2      ......
 3      if ( dst . distance ≤ delta ){
 4          Respawn ( dst )
 5      else
 6          WL.push ( dst )
 7      ......
 8      ......
 9      }
10  }
11  Pipe Once {
12      Invoke INIT( graph , src )
13      Pipe {
14          Invoke SSSP( graph, curdelta ) ;
15          Invoke remove_dups( ) ;
16          curdelta += DELTA ;
17      }
18  }
```

4.4.2.2 Lonestar-GPU

The LonestarGPU [122] framework supports mutation of graph objects and implementation of cautious morph algorithms. It has implementations of several cautious morph algorithms like Delaunay Mesh Refinement, Survey Propagation, Boruvka's-MST and Points-to-Analysis. Boruvka's-MST algorithm implementation uses the `Union-Find` data structure. LonestarGPU also has implementations of algorithms like SSSP, BFS, Connected Components etc., with and without using *worklists*. LonestarGPU does not provide any API based programming style.

4.4.2.3 MapGraph

MapGraph [123] is an open source framework which uses the vertex centric *Gather-Apply-Scatter* model of execution and targets single GPUs. It uses compressed sparse row (CSR) format to store graph objects. MapGraph decides the scheduling policy at runtime based on the number of active vertices and the size of the adjacency lists for the active vertices at each superstep.

The *Apply* phase in the GAS execution model is parallel and well optimized. The MapGraph framework optimizes the *Gather* and *Scatter* phases of the GAS execution model. It uses *dynamic scheduling* and *two-phase decomposition* to achieve the same. Dynamic scheduling uses the *CTA-based* scheduling strategy based on the degree of an active vertex. An active vertex is assigned to a CTA,[4] and each thread of the CTA processes only one neighbor of the vertex when the vertex degree is large (like in social network graphs). *Scan-based* scattering uses a *prefix sum* operation to compute the starting and ending points in the column-indices array. Then an entire CTA is assigned to gather the referenced neighbors from the column-indices array using the scatter vector. *Warp-based* scattering does a coarse-grained redistribution of the scattering workloads. These three policies are used to obtain better *load balance* and improved *memory access* patterns. Dynamic scheduling is very efficient for BFS and SSSP. But it may lead to imbalanced workloads among CTAs.

4.4.2.4   Other Frameworks

The Puffin [124] framework proposes a novel data representation, which meets programming needs with minimum storage space requirements. The framework overlaps communication and computation using novel runtime strategies. The runtime system of Puffin divides the tasks and manages the order of execution of different kinds of tasks. Puffin provides both vertex-centric and edge-centric programming models that are used in graph analytics. Tigr [125] transforms graphs for GPU-friendly computation. Tigr Transforms irregular graphs to more regular ones by changing the topology without graph partitioning. The *split transformation* of Tigr splits vertices with very high degree iteratively until the degree of each vertex is within a predefined limit. The *virtual split transformation* of Tigr adds a virtual layer on top of the input graph. The virtualization separates the programming abstraction from the input graph. The computation tasks are scheduled at the virtual layer of the transformed graph. The actual value propagation is carried to the input graph. GraphBLAS extension for GPUs is called GraphBLAS Template Library (GBTL). An effort to standardize GraphBLAS for GPU is reported in [126]. A few works [127, 128] also focus on graph analytics on a single GPU.

4.4.3   Multi GPU

4.4.3.1   Medusa

Medusa [129] is a programming framework for graph algorithms on GPUs and multi-GPU devices. It provides a set of API and a runtime system for this purpose. A

---

[4]Cooperative Thread Array.

Table 4.1 Medusa API

APIType	Parameter	Variant	Description
ELIST	Vertex V, Edgelist el	Collective	Apply to edge-list el of each vertex v
EDGE	Edge e	Individual	Apply to each edge e
MLIST	Vertex v, Message-list ml	Collective	Apply to message-list ml of each vertex v
MESSAGE	Message m	Individual	Apply to each message m
VERTEX	Vertex v	Individual	Apply to each vertex v
Combiner	Associative operation	Collective	Apply an associative operation to all edge-lists or message-lists

programmer is required to write only sequential $C++$ code with these API. Medusa provides a programming model called the *Edge-Message-Vertex* or *EMV* model. It provides API for processing vertices, edges or messages on GPUs. A programmer can implement an algorithm using these API. Medusa API are presented in Table 4.1. API on vertices and edges can also send messages to neighbouring vertices. Medusa programs require user-defined data structures and implementation of Medusa API for an algorithm. The Medusa framework automatically converts the Medusa API code into CUDA code. The API of Medusa hide most of the CUDA specific details. The generated CUDA code is then compiled and linked with Medusa libraries. Medusa runtime system is responsible for running programmer-written codes (with Medusa API) in parallel on GPUs.

Algorithm 4.12 shows the pagerank algorithm implementation using Medusa API. The pagerank algorithm is defined in Lines 26–31. It consists of three user-defined API: *SendRank* (Lines 2–7) which operates on *EdgeList*, a vertex API *UpdateVertex* (Lines 9–13) which operates over the vertices and a *Combiner()* function. The *Combiner()* function is for combining message values received from the *Edgelist* operator, which sends messages using the *sendMsg* function (Line 6). The *Combiner()* operation type is defined as addition (Line 36) and message type as float (Line 37) in the *main()* function. The *main()* function also defines the number of iterations for *pagerank()* function as 30 (Line 39) and then the *pagerank()* function is called using *Medusa::Run()* (Line 41). The *main()* function in Medusa code initializes algorithm-specific parameters like *msgtype*, *aggregator* function, number of GPUs, number of iterations etc. It then loads the graph onto GPU(s) and calls the *Medusa::Run* function which consists of the main kernel. After the kernel finishes its execution, the result is copied using the *Dump_Result* function (Line 42).

The *SendRank* API takes an *EdgeList el* and a *vertex v* as arguments, and computes a new value for *v.rank*. This value is sent to all the neighbours of the vertex *v* stored in *Edgelist el*. The value sent using the *sendMsg* function is then aggregated using the *Combiner()* function (Line 29) which is defined as the *sum* of the values received. The *UpdateVertex Vertex* API then updates the pagerank using the standard equation to compute the pagerank of a vertex (Line 12).

Algorithm 4.12: Medusa Pagerank Algorithm

```
 1  // Device code API:
 2  struct SendRank{ // ELIST API
 3  __device__ void operator() (EdgeList el, Vertex v) {
 4      int edge_count = v.edge_count;
 5      float msg = v.rank/edge_count;
 6      for(int i = 0; i <edge_count; i ++) el[i].sendMsg(msg);
 7  }
 8  };
 9  struct UpdateVertex{ // VERTEX API
10  __device__ void operator() (Vertex v, int super_step) {
11      float msg_sum = v.combined_msg();
12      vertex.rank = 0.15 + msg_sum*0.85;
13  }
14  };
15  struct vertex{ //Data structure definitions:
16  float pg_value;
17  int vertex_id;
18  };
19  struct edge{
20  int head_vertex_id, tail_vertex_id;
21  };
22  struct message{
23  float pg_value;
24  };
25  Iteration definition:
26  void PageRank() {
27      InitMessageBuffer(0); /* Initiate message buffer to 0 */
28      EMV<ELIST>::Run(SendRank); /* Invoke the ELIST API */
29      Combiner(); /* Invoke the message combiner */
30      EMV<VERTEX>::Run(UpdateRank); /* Invoke the VERTEX API */
31  }
32  int main(int argc, char **argv) {
33      ......
34      Graph my_graph;
35      // load the input graph.
36      conf.combinerOpType = MEDUSA_SUM;
37      conf.combinerDataType = MEDUSA_FLOAT;
38      conf.gpuCount = 1;
39      conf.maxIteration = 30;
40      Init_Device_DS(my_graph); /*Setup device data structure.*/
41      Medusa::Run(PageRank);
42      Dump_Result(my_graph);/* Retrieve results to my_graph. */
43      ......
44      return 0;
45  }
```

Medusa supports execution of graph algorithms on multiple GPUs of the same machine, by partitioning a large input graph and storing the partitions on multiple GPUs. Medusa uses the EMV model (Edge-Message-Vertex), which is an extension of the Bulk Synchronous Parallel (BSP) Model. Medusa does not support running different algorithms on different devices at the same, when a graph object fits within a single GPU. Also, it does not support distributed execution on GPU clusters.

4.4.3.2 Lux

Lux [91] is a framework targeting distributed multi-GPU systems. It favors iterative graph algorithms. Its approach is implicit in many frameworks that have already been discussed and a few of them are [11, 97, 129]. Lux is implemented on top of Realm [130]. The *in-degree* and *out-degree* of a vertex v are denoted by $deg^-(v)$ and $deg^+(v)$ respectively in Lux. The *innbrs* and *outnbrs* of a vertex are denoted by $N^-(v)$ and $N^+(v)$ respectively. Graph processing in Lux is stateless and needs to be implemented on the Lux API shown in Algorithm 4.13. By implementing the interface functions, a computation is factored into the *init, compute*, and *update* functions. This is similar to the other iterative approaches such as *Gather-Apply-Scatter* and *Advance-Filter-Compute*. Lux supports *push* and *pull* based computation models.

Algorithm 4.13: Lux Framework API

```
1  interface Program (V, E) {
2      void init (Vertex v, Vertex v_old);
3      void compute (Vertex v, Vertex u_old, Edge e);
4      bool update (Vertex v, Vertex v_old);
5  }
```

Lux initializes the vertex properties for an iteration by running the *init()* function on all the vertices. The properties of the vertices from the previous iteration (v_{old}) are passed as immutable inputs to *init()* function. The *compute()* function takes an edge *e(u,v)* and its properties, and the properties of the vertex *u* from the previous iteration (u_{old}) as inputs and updates the properties of the vertex *v*. The input properties are immutable in the *compute()* function. The order of processing the edges is non-deterministic. The *compute* function is executed concurrently on many vertices. As the last step of an iteration, the *update()* function is called on every vertex and updates are committed. The Lux runtime exits when no vertex properties are updated in an iteration, denoting a fixed point. Algorithms 4.14 and 4.15 present examples of programming the Pagerank algorithm in Lux.

Lux is a distributed framework. It partitions a large graph to subgraphs for different compute nodes. Lux stores the vertex updates in the zero-copy memory. The zero-copy memory is shared among all GPUs on a compute node. The GPUs store mutable vertex properties in the shared zero-copy memory, which can be directly loaded by other GPUs. It makes use of the GPU shared memory to store subgraphs. The *partially-shared* design of Lux reduces the memory for vertex updates and substantially reduces the communication between compute nodes. Lux uses *edge-cut* partitioning to create subgraphs with almost equal number of edges in each subgraph. It guarantees coalesced accesses using its partitioning method, which increases the throughput. Lux uses compressed sparse row (CSR) format to store graphs, and its runtime performs fast *dynamic repartitioning* which achieves efficient load-balancing. Workload imbalance is detected by monitoring runtime performance and then recomputing the partitioning in order to restore balance.

Algorithm 4.14: Lux: PageRank Using Pull Model

```
1  define Vertex { rank: float }
2  void init ( Vertex v, Vertex vold ) {
3  |   v.rank = 0;
4  }
5  void compute ( Vertex v, Vertex uold, Edge e ) {
6  |   atomicAdd (& v.rank, uold.rank );
7  }
8  bool update ( Vertex v, Vertex vold ) {
9  |   v.rank = (1 - d ) / |V| + d * v.rank;
10 |   v.rank = v.rank / deg+(v);
11 |   return (|v.rank - vold.rank| > δ);
12 }
```

$v.rank = (1 - d) / |V| + d * v.rank$

$v.rank = v.rank / deg^+(v)$

$return\ (|v.rank - v_{old}.rank| > \delta)$

Algorithm 4.15: Lux: PageRank Using Push Model

```
1  define Vertex { rank, delta: float }
2  void init ( Vertex v, Vertex vold ) {
3  |   v.delta = 0;
4  }
5  void compute(Vertex v, Vertex uold, Edge e) {
6  |   atomicAdd(&v.delta, uold.delta);
7  }
8  bool update(Vertex v, Vertex vold) {
9  |   v.rank = vold.rank+ d * v.delta
10 |   v.delta = d * v.delta / deg+(v)
11 |   return (|v.delta| > δ)
12 }
```

$v.rank = v_{old}.rank + d * v.delta$

$v.delta = d * v.delta / deg^+(v)$

$return\ (|v.delta| > \delta)$

4.5 Frameworks for Distributed Systems

Partitioning methods for distributed graph analytics have been compared in [131]. Different policies include *balanced vertices*, *balanced edges* and *balanced vertices and edges*. The 2D block partitioning uses the adjacency matrix which is divided into blocks in both dimensions, and assigns one or more blocks to each node. CheckerBoard Vertex-Cuts (BVC), Cartesian Vertex-Cuts (CVC) and Jagged Vertex-Cuts (JVC) are examples of 2D partitioning. *Vertex-cut* and *edge-cut* partitioning methods have already been discussed in Sect. 1.4. Distributed execution involves computation across different compute nodes in parallel, and communication between nodes to exchange information. Efficiency is achieved if there is load balance and less communication overhead. As the problem of optimal graph partitioning is NP-hard, we rely on heuristics. We compare different distributed graph analytics frameworks in this section.

4.5.1 Distributed System with CPU

GoFFish [132] follows a modified form of the vertex-centric execution model of Pregel. Primarily, it provides the API functions listed below:

* *computation—Compute(Subgraph, Iterator<Message>)*
  The *Compute*() method processes a subgraph stored on a computing node. The vertices and edges can have mutable properties but the graph topology does not change. The function is meant to fully traverse the subgraph upto the remote edges within a single superstep and update mutable properties of the local vertices and edges.
* *message-passing*—The subgraphs of two different nodes communicate with each other using *message-passing*, with messages exchanged in each superstep boundary. *SendToAllSubGraphNeighbors*() sends messages to all the neighbouring subgraphs (nodes). As other subgraphs are discovered across supersteps, the functions *SendToSubGraph*() and *SendToSubGraphVertex*() functions are used for communication. The function *SendToAllSubGraphs*() broadcasts to all subgraphs, which is a very expensive operation.

The *Compute( )* method can invoke *VoteToHalt( )*, as done in Pregel. The application terminates when there are no messages sent by any subgraph in a superstep.

The GraphX [133] framework provides two separate API: a variant of the Pregel API and a Map-Reduce style API. It is implemented on top of the Apache Spark.[5] GraphX allows *composition* of graphs with unstructured and tabular data. The same data can be viewed as both graph and as collections without data replication and movement. Thus, it allows users to adopt either graph or collection computation pattern, that is best suited for a given application with the same performance. The vertex-cut partitioning method is used to divide graphs as horizontally partitioned collections. It provides a distributed join and *view optimizations*, enabling high throughput. A graph is represented as a *property graph* with each vertex and edge having properties which may include metadata. A *property graph* can also be

---

[5]https://spark.apache.org.

represented as vertex and edge property collections. The *vertex collection* properties are keyed by vertex identifiers.

The Zorro [134] system provides efficient fault tolerance for distributed graph analytics frameworks. Zorro was integrated into PowerGraph. Zorro opportunistically exploits vertex replication available in the framework to rebuild the state of failed servers. The cost involved in rebuilding the system is very less. The Zorro system quickly recovers over 99% of the graph state when a few servers fail, and between 87–92% when half of the cluster fails. Zorro produces results with little to zero inaccuracy in experiments during failure of cluster nodes.

PGX.D [135] uses a fast cooperative context-switching mechanism. It has a low-overhead, bandwidth-efficient communication framework with remote data-pulling patterns. It reduces communication volume and provides workload balance by applying selective ghost nodes, edge partitioning, and edge chunking transparently.

There are few other distributed graph analytics frameworks for CPU clusters such as [136–138].

4.5.2   Distributed System with CPU and GPU

Gluon [139] is a communication optimizing framework for graph analytics on distributed systems with CPU and GPU. The Gluon framework has inbuilt partitioning policies and a programming model. It has its own API and a shared memory model. The programmer is required to implement the algorithmic logic within the Gluon API. Gluon then converts the programs to run on distributed heterogeneous systems with CPU and GPU. The Gluon framework minimizes the communication overhead by looking at the structural and temporal properties of graph partitioning. Gluon is integrated with Galois and Ligra which are the shared memory frameworks for CPU and thus makes available, distributed versions of Galois (D-Galois) and Ligra (D-Ligra). Gluon is interfaced with the single GPU framework IrGL, to produce a distributed memory GPU framework (D-IrGL). Gluon provides different partitioning methods and optimizes the communication overhead. While performing distributed execution, the Gluon framework uses Galois or Ligra framework on CPU, and uses IrGL framework on GPU. When Gluon is executed on a CPU cluster, it is either D-Galois or D-Ligra.

Gluon was extended to have fault tolerance in Phoenix [140]. The computation restarts from a state which will finally produce correct results. This is not in accordance with the traditional approach where computation happens from a previous safe state. Phoenix has an API where the programmer can specify the state adjustment. There is no overhead when Phoenix performs normal (fault-free) execution. The novelty of Phoenix is its low overhead and capability to produce correct results after fault recovery. Typically, a framework may have either high overhead or compute approximate result on fault recovery. The overhead is often due to the presence of checkpointing for fault recovery in distributed graph analytics [11, 17]. Checkpoing adds overheads even in the absence of faults.

Confined recovery has been adopted in distributed frameworks for CPUs where rollback is not required on fault [109, 133].

Gluon and Phoenix follow the BSP execution model. Gluon-Async extends Gluon for asynchronous execution in distributed heterogeneous graph analytics [141]. The execution model of Gluon-Async is lock-free, non-blocking, and asynchronous, and is named bulk-asynchronous parallel (BASP). It combines the advantages of BSP models and asynchronous execution. The computation happens in supersteps. The individual hosts do not wait for the completion of the round from other hosts. Each host sends messages and receives available messages and moves to the next round.

Chapter 5
GPU Architecture and Programming Challenges

This chapter provides an overview of GPU architectures and CUDA programming. The performance of the same graph algorithms on multi-core CPU and GPU are usually very different. Intricacies of thread scheduling, barrier synchronization, warp based execution, memory hierarchy, and their effects on graph analytics are illustrated with simple examples.

5.1 Introduction

Specialized software running on a general purpose computer may not give required performance for different application domains. A general purpose computer may not be able to harness the parallelism available in the software due to several reasons such as lack of sufficient number of CPU function units, ability to process vectors, slow synchronization, slow memory management, etc. This issue led to the design of *hardware accelerators*. Hardware acceleration is implemented with a dedicated hardware for a specific domain or application. Memory management unit (MMU) and floating point arithmetic co-processor, are examples for elementary hardware accelerators. Graphical Processing Units (GPUs), Field Programmable Gate Array (FPGA) and Application Specific Integrated Circuits (ASICs) are examples for complex hardware accelerators. Many application domains benefit from hardware acceleration. Table 5.1 describes available hardware accelerators for different application domains. Hardware acceleration is very critical to achieve high performance. Dedicated hardware is energy-efficient as it consumes less power compared to a software based solution on a general purpose computer.

Flynn has classified the computing architecture [142] into four classes mentioned below:

- Single Instruction Single Data (SISD)
- Single Instruction Multiple Data (SIMD)

© Springer Nature Switzerland AG 2020
U. Cheramangalath et al., *Distributed Graph Analytics*,
https://doi.org/10.1007/978-3-030-41886-1_5

Table 5.1 Hardware acceleration in different domains

Domain	Hardware accelerators	Commercial availability examples
Computer graphics	Graphics processing unit (GPU)	Nvidia, Intel, AMD
Digital signal processing	Digital signal processor (DSP)	Texas Instruments, Analog Devices
Computer networks	Network-on-chip (NoC)	Arteris, NetSpeed Systems
Artificial intelligence	Vision processing unit (VPU)	Myriad X (Intel)
		Pixel visual core by Google
Multilinear algebra	Tensor processing unit (TPU)	TPU by Google

- Multiple Instruction Single Data (MISD)
- Multiple Instruction Multiple Data (MIMD)

SISD processors come with a sequential architecture (single-core) with parallelism limited among and between instructions such as *software pipelining* and *instruction reordering*. An SIMD architecture has many cores and provides parallelism where all the computing cores perform the same operation (instruction) on different data elements. SIMD operations are available in vector processors which have very large instruction width (VLIW processors). GPU devices follow Single Instruction Multiple Thread (SIMT) architecture where groups of threads (called a *warp*) execute the same instruction at the same point of time. Different *warps* may execute different instructions at the same time. MISD architecture is not common. Multiple instructions operate on different data elements in the MIMD architecture. A multi-core CPU is an example of an MIMD architecture. Each core is a CPU with a separate program counter and performs operations as dictated by its program counter.

5.2   Graphics Processing Unit (GPU)

A *massively parallel* GPU device is of particular interest to all application domains. GPU devices have very high computing power and energy efficiency. About a decade ago, GPUs were mostly used as graphics co-processors in workstations, but now, general purpose computing on GPUs is popular in addition to its usage for computer graphics.

5.2.1   GPU Architecture

GPU devices follow the SIMT architecture. GPU devices provide *massive-parallelism* with a huge (typically more than thousand) number of cores which are hierarchically clustered. GPU devices are attached as accelerators to host multi-core CPUs (main CPU of the computing unit), and data needs to be copied between the volatile memory of CPU and GPU. A GPU device cannot directly access the nonvolatile memory or secondary storage of the host device. A GPU device consists

Table 5.2 Comparison of different GPU micro-architectures

Name	Micro-architecture	Total cores (SPs)	Core frequency	Memory size (type)	GFLOPS	Release year
GK-110B	Tesla (K40)	2880	745 MHZ	12 GB (GDDR5)	4290	2013
GTX-870M	Kepler	1344	941 MHz	6 GB (GDDR5)	3050	2014
GTX-1080	Pascal	2560	1.607 GHz	8 GB (GDDR5)	10,600	2017
GV-100	Volta	5120	1.2 GHz	32 GB (HBM2)	14,800	2018

of multiple Streaming Multiprocessors (SMs), and each SM consists of many Streaming Processors (SPs). The number of SPs in an SM is a multiple of the warp size which is usually thirty two. All the SPs in the same warp always follow the same control flow and execute the same instruction at the same time. An SP usually runs a thread.

Table 5.2 shows the important properties of the recent Nvidia GPUs. Each GPU in the table is from a different micro-architecture family, namely Tesla [143], Kepler [144], Pascal [145], and Volta [146]. The columns in the table show the GPU name, its micro-architecture, total numbers of SPs in the GPU, frequency of the SP in MHz, the volatile global memory size of the GPU in GB, single precision floating point operations per second (FLOPS) as a multiple of 10^6 (Giga), and the release date of the GPU respectively. The frequency of the SPs in a GPU and the volatile memory of a GPU are lesser than that of multi-core CPUs of the same time. However, the optimal computation power of a GPU is much higher than that of a multi-core CPU. For example, the optimal performance of the *GV-100* GPU is 14,800 GFLOPS for single precision floating point operations while that of Intel *Core i7-6700K* CPU is only 114 GFLOPS. *Core i7-6700K* CPU supports frequencies up to 4 GHz and a volatile memory of more than 100 GB. The optimal performance on a GPU device depends on many factors such as amount of parallelism in the computation, *warp divergence*, number of atomic operations, amount of communication between host CPU and GPU during the computation etc.

5.2.2 Multi-GPU Machine

A general purpose computer with multiple GPUs attached to it is known as a *multi-GPU machine*. The communication overhead between two GPUs depends on how the GPU devices are attached to the I/O Hub (IOH) and how the program is written. Multi-GPU machines can be used to perform parallel graph analytics. The computations can be done independently, in parallel on each GPU. This can be achieved by running different algorithms on different GPUs, or the same algorithm with different input graph objects on different GPUs. This is possible when each input graph object fits within the volatile memory of the concerned single GPU.

A multi-GPU machine is useful for distributed computation when a graph object does not fit within the volatile memory of a single GPU. If the GPUs are connected

---

Algorithm 5.1: Enable PeerAccess Between GPUs on the Same IOH

```
 1  int *peerarray;
 2  int enablepeeraccess() {
 3      int devCount = 0,flag = 0;
 4      cudaGetDeviceCount(&devCount); // devcount stores number of GPUs
 5      if(devCount == 0)return -1; // No GPUs present
 6      peerarray = (int *)malloc(sizeof(int) * devCount * devCount);
 7      for int i = 0; i < devCount; ++i {
 8          for (int j = i + 1; j < devCount; j++) {
 9              flag = 0; // check peeraccess between device i and j i≠j
10              cudaSetDevice(i);
11              if (cudaDeviceEnablePeerAccess(j, 0) != cudaSuccess) flag = 1;
12              cudaSetDevice(j);
13              if (cudaDeviceEnablePeerAccess(i, 0) != cudaSuccess) flag = 1;
14              if(flag == 0) peerarray[i * devCount + j] = peerarray[j * devCount + i] = 1;
15          }
16      }
17      return 0;
18  }
```

---

to different IOHs, then the data will have to go through the system memory and the communication overhead is high. Otherwise, DMA transfer is possible between the GPUs connected to the same IOH, and the overhead is lower. If two GPUs G_1 and G_2 are connected to the same IOH, G_1 (G_2) can read or write to the memory of G_2 $(G_1$ respectively). This needs to be enabled by calling the API functions (provided by CUDA) before the communication happens. DMA transfer between GPUs on the same IOH is known as *peer-access*. A sample CUDA code to enable *peer-access* between GPUs on a machine is shown in Algorithm 5.1. The function call (*cudaDeviceEnablePeerAccess()*) in the program will be successful (i.e., returns cudaSuccess) if the GPUs are connected on the same IOH. Then onwards, communication between GPUs on the same IOH will happen using DMA. PeerAccess and normal access in a multi-GPU machine is shown in Fig. 5.1a, b respectively.

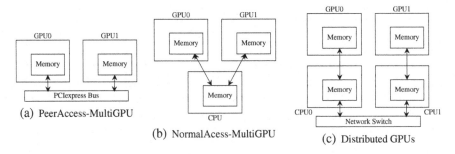

(a) PeerAccess-MultiGPU

(b) NormalAcess-MultiGPU

(c) Distributed GPUs

Fig. 5.1 Communication in multi-GPU machines and GPU clusters

5.2.3   Distributed GPU Systems

A *GPU cluster* consists of a collection of separate single or multi-GPU machines connected to each other using a network switch or hub. Distributed computation using a GPU cluster has high communication overhead. The data transfer between two GPUs present in the cluster is very expensive as the data must hop from one device to another before reaching its destination (see Fig. 5.1c). The communication overhead in a distributed GPU cluster is very high compared to the computational speed of GPUs. Effective performance of a GPU cluster is not very good for graph analytics application domains due to this high communication overhead. Figure 5.1 shows the interconnection for two GPU systems which are connected in different ways. Similar connection is possible for more than two GPUs.

5.3   General Purpose Computing on GPU (GPGPU)

GPUs have a massively parallel architecture with very high computing power compared to that of multi-core CPUs, but the amount of volatile memory available in a GPU (tens of GB) is much lesser than that in a multi-core CPU (hundreds of GB). The high computing power of GPUs is utilized for general purpose computing in various domains. Prior art comparing relative performance of GPUs and multi-core CPUs has concluded that GPU devices outperform multi-core CPUs in application domains such as graph analytics [26], machine learning [147], image processing [148], to name a few. The performance of graph anaytics applications on a GPU depends on graph properties such as *diameter* and *degree distribution* of the graph object. General purpose computing on GPU is very efficient for benchmarks when there is minimal communication between a *device* and its *host* during the computation. Otherwise, there is significant runtime overhead and the performance suffers. Such processing may require a special hardware architecture (*hardware acceleration*) to improve the performance [149]. A typical example for such a hardware accelerator is an *integrated GPU* which reduces the communication overhead between CPU and GPU.

5.3.1   CUDA Programming

The performance of Nvidia GPUs are best harnessed when programmed with CUDA. A CUDA programmer is required to take care of memory management, data transfer between CPU and GPU, thread management etc. Programming in CUDA is therefore even more challenging than programming in high-level languages such as C or C++. The CUDA programming model follows the syntax of C++ with additional keywords and library functions. The keywords and functions have been

Table 5.3 CUDA—important keywords and functions for CPU (*host*) and GPU (*device*)

Item	Type	Description
__global__	Keyword	Function run on *device*, called from *host*
__device__	Keyword	Function or variable accessible from *device*
__syncthreads()	Keyword	Barrier for a thread block.
cudaMemcpy	Function	Transfer Data between *host* and *device*
cudaMemcpyFromSymbol	Function	Copy data from *device* to *host*
cudaMemcpyToSymbol	Function	Copy data from *host* to *device*
cudaMalloc	Function	Allocate memory on *device*
cudaDeviceSynchronize	Function	Barrier for the CUDA kernel (called from *host*)

added to differentiate between CPU aspects and GPU aspects, such as, the CPU memory and the GPU memory, functions executed on the CPU and the GPU, etc. The library functions for thread and memory management are also present in CUDA. Table 5.3 lists some of the important keywords and functions in CUDA. There are CUDA libraries such as *Thrust* [4] and CUB[1] which provide STL-like high-level abstractions for GPGPU. Good performance is not always guaranteed when special libraries are used for computation.

5.3.2 GPU Thread Scheduling

A parallel *device* (GPU) function called from the *host* (CPU) is called a CUDA kernel. A CUDA kernel is called with *thread blocks* and *threads-per-block*, with both having three dimensions *x*, *y*, and *z*. One dimensional indexing, where values are specified only for the *x* dimension, is often used in practice. A one dimensional grid consisting of *p* number of thread blocks and with *q* number of threads-per-block, will have *p* × *q* number of threads running the CUDA kernel. Each kernel thread is identified by a unique identifier which is calculated by the formula given below (shown for one dimensional grid for simplicity).

$$tid = blockIdx.x \times blockDimx.x + threadIdx.x$$

Here, the variables *blockIdx.x* and *threadIdx.x* represent the thread block number and the thread number within the thread block respectively, *blockDimx.x* and *gridDim.x* represent the number of threads in a block, and the number of thread blocks in a kernel respectively. There is a limit on the maximum number of thread blocks and threads in a thread block specific to each device. GPU devices with

---

[1] https://nvlabs.github.io/cub/.

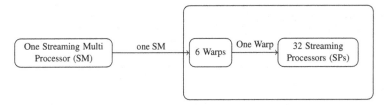

Fig. 5.2 NVidia GPU with—each SM having 192 SPs (six warps—6 × 32)

compute capability[2] greater than two have the maximum number of threads in a thread block set to 1024. The number of thread blocks differs with GPU devices, with values such as 64×1024 thread blocks or more per kernel possible. In Tesla K20 GPU, the maximum dimension values for a thread block are (1024, 1024, 64) and the maximum dimension size for a grid is (2147483647, 65535, 65535) for the triple (x, y, z).

The GPU device computing core hierarchy is shown in Fig. 5.2. The GPU device consist of $(r = p \times q)$ number of streaming processors (SPs), where p is the number of streaming multiprocessors(SMs), and q is the number of SPs in each SM. The individual thread blocks of a CUDA kernel are assigned to an SM by the CUDA *warp-scheduler*.[3] The threads in the thread block are executed as a collection of warps (thirty two threads) in SIMT fashion. Once a thread block is assigned to an SM, the thread block cannot migrate to a different SM. The CUDA runtime scheduler will assign another thread block to an SM once the SM has completed the computation of the previously assigned thread block. The process repeats until all the thread blocks of a CUDA kernel have executed on various SMs in the order decided by the CUDA runtime scheduler.

CUDA kernel calls are *asynchronous*. This means, after launching a kernel, the next CPU code continues to execute while the GPU kernel is being executed concurrently. The *cudaDeviceSynchronize()* function call needs to be added in the source program after the kernel call for *synchronous* execution (alternatively, one may use events or synchronize using waiting loops). The *cudaDeviceSynchronize()* function call acts as a barrier for the *host*, and the *host* waits for completion of the kernel execution on the *device* before going to the next instruction.

The CUDA runtime supports a barrier for threads within a single thread block using the library function *__syncthreads()*. A programmer is required to place the *__syncthreads()* function call at a proper place so that all the threads within a thread block execute the function call. If it is placed inside a conditional code, all the threads may not execute the function call. This may lead to threads not meeting at a program point, resulting in a hung-kernel.

---

[2]The compute capability is the "feature set" (both hardware and software features) of the device. As devices and CUDA versions get more and more features, the compute capability also increases.

[3]A warp is just a group of threads and its size is set by the hardware, usually as 32.

Algorithm 5.2: Barrier for CUDA Kernel

```
1   __device__ void __gpu_global_barrier(int val, volatile int *A_in) {
2       int tid = threadIdx.x; // thread id in block
3       int numblocks = gridDim.x; // number of blocks
4       int bid = blockIdx.x; //block id
5       if (tid == 0) A_in[bid] = val; // thread zero sets A_in[bid]=Val;
6       // block zero waits for A_in[0] to A_in[numblocks-1] to become Val
7       if( (bid == 0) ){
8           if( (tid < nBlockNum) ){
9               while( (A_in[tid] != val) ){
10                  // loop until all the blocks sets A_in[bid] to val
11              }
12          }
13          __syncthreads();// executed by all threads of thread block zero
14      }
15      __syncthreads();// executed by all threads in CUDA Kernel.
16  }
```

The CUDA runtime does not support a *global-barrier* for all the threads of a kernel. This is in contrast to the multi-core parallel libraries (e.g., OpenMP), which provide such global barriers. A global-barrier for a CUDA kernel needs to be implemented in software. A lock-free software implementation of the global barrier has been devised [117]. A requirement in the implementation of a *global-barrier* is that a CUDA kernel must be called with the number of thread blocks less than or equal to the number of SMs on the particular GPU. This will require the CUDA kernel to be called multiple times from the *host*, as the host calls the kernel with fewer threads. However, this is necessary to prevent deadlocks. Imagine a situation in which several thread blocks are mapped to a single SM and the active block has a global barrier, over which it waits for completion. This leads to a deadlock since the active thread block cannot be preempted and it does not exit without the other thread blocks reaching the barrier. Other thread blocks cannot execute without preemption of the active block.

A sample code for a CUDA kernel barrier is shown in Algorithm 5.2. The basic idea is to assign a synchronization variable to each thread block. Thread block i will be assigned location $A_{in}[i]$, and thread zero of thread block i will be responsible for setting $A_{in}[i]$. N threads of thread block zero (N is the number of thread blocks) busy-wait on A_{in}, thread j checking $A_{in}[j]$. The *__syncthreads()* call at Line 13 ensures synchronization of all the threads in thread block zero.

5.3.3 Warp Based Execution

Algorithm 5.3 contains a CUDA kernel named *kernel* (Lines 4–14) and a *device* function *devfun* (Lines 1–3) called from *kernel*. Consider a GPU device with 192 cores per SM and a CUDA kernel launched with 960 threads-per-block. The *warp-scheduler* assigns a thread block of 960 threads to each SM (assuming that the number of total threads is more than the total number of SPs in the GPU). The threads in a thread block are executed as a collection of warps (32 threads per warp) which follow the SIMT architecture. The number of warps in the thread block is thirty (960÷32). An SM has only 192 cores and at a time only six (192÷32) warps can run on an SM. A new warp is possibly loaded into the SM upon termination of a running warp, since usually, there is no preemption of a warp. The thread block finishes execution when all the thirty warps finish execution.

Algorithm 5.3: Warp Based Execution-Example

```
 1   __device__ devfun() {
 2   |   // perform some computation
 3   }
 4   __global__ kernel(int size, int *A, bool *B) {
 5   |   int id = blockIdx.x * blockDim.x + threadIdx.x;
 6   |   if( (id < size && B[id] == true) ){
 7   |   |   int cnt = A[id];
 8   |   |   for (int i = 0; i < cnt; i++) {
 9   |   |   |   ....
10   |   |   |   devfun();
11   |   |   |   .....
12   |   |   }
13   |   }
14   }
```

The threads within a single warp follow the same control flow, unlike in the MIMD model followed on a multi-core CPU. The program in Algorithm 5.3 contains a conditional block (see Lines 6–13). The CUDA compiler produces predicated instructions for the true and false blocks of statements. All the instructions will be executed but for those instructions for which the predicate is false, the effect is that of executing a NULL (or NOP) instruction. In the current example, even if one thread in a warp satisfies the condition ($id < size \&\& B[id] == true$), all the warp-threads enter the conditional block, but except one, the others will not do anything useful. In general, if some threads satisfy a condition and the others do not, all the threads end up executing both the true and the false code blocks, with loss of performance. This is known as *warp divergence*. The best situation is when all the warp-threads satisfy the condition, or when none of the warp-threads satisfies the condition. In such a case, all the threads execute only one block of instructions and overheads of warp divergence are zero. With large nested if-then blocks, overheads of warp divergence (execution of dummy instructions) can reduce performance upto 32 times. The lack of *warp divergence* in a CUDA kernel is a necessary (not sufficient) condition for optimal throughput. Conditional blocks are the primary candidates for *warp divergence*. Good performance also demands

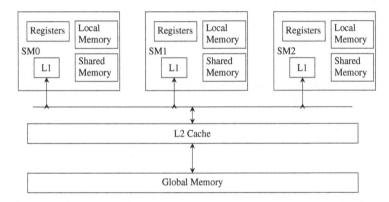

Fig. 5.3 GPU memory hierarchy

warps to have *coalesced* memory access patterns. Scattered access patterns result in difference in access latency among threads which increases the execution time.

5.3.4   Memory Hierarchy

GPU memory hierarchy is shown in Fig. 5.3. The SMs in the GPU consist of registers, local memory, shared memory, and L1 cache. The shared memory should be used effectively by a CUDA kernel for throughput. The program in Algorithm 5.4 shows the computation of sum of all the elements in an array, where the number of threads in the thread block is 1024. As many thread blocks as necessary to take care of the whole array *ptr* are launched in the main program (see Line 23).

The addition of values of elements for each thread block is carried out using the GPU shared memory. Each thread initializes one location of the shared array *blockcount* in Line 7. Summing of the elements of the shared array *blockcount* takes place in the loop (see Line 10), which makes $log_2(1024) = 10$ iterations. In the first iteration ($i = 2$), thread 0 sums up *blockcount*[0] and *blockcount*[1], thread 2 sums up *blockcount*[2] and *blockcount*[3], thread 4 sums up *blockcount*[4] and *blockcount*[5], and so on, and leave the results in *blockcount*[0], *blockcount*[2], *blockcount*[4], etc., respectively. In the second iteration ($i = 4$), threads 0, 4, 8, . . ., only are active. Thread 0 sums up *blockcount*[0] and *blockcount*[2], thread 4 sums up *blockcount*[4] and *blockcount*[6], thread 8 sums up *blockcount*[8] and *blockcount*[10], etc., and leave the results in *blockcount*[0], *blockcount*[4], *block-count*[8], etc., respectively. Similarly, in the third iteration ($i = 8$), sum of the elements *blockcount*[0..7] is computed into *blockcount*[0], sum of the elements *blockcount*[8..15] is computed into *blockcount*[8], etc. Finally, the sum of all the elements of *blockcount* will be available in *blockcount*[0].

The barrier for a thread block is achieved using __*syncthreads()* library function. The local value computed by an SM for each thread block is added *atomically* to the

global device memory location for the variable *reduxsum*. The above computation
reduces the number of atomic operations and global memory accesses. This results
in good performance and throughput.

Algorithm 5.4: Reduction Operation Using Shared Memory—CUDA

```
1   __device__ int reduxsum;
2   __global__ void findsum(int *arr, int size) {
3       int id = blockIdx.x * blockDim.x + threadIdx.x;
4       __shared__ volatile unsigned int blockcount[1024];
5       blockcount[threadIdx.x] = 0;
6       if( (id < size) ){
7           blockcount[threadIdx.x]= arr[id];
8       }
9       __syncthreads();
10      for (int i = 2;i <= 1024; i = i * 2) {
11          if (threadIdx.x % i == 0) blockcount[threadIdx.x] += blockcount[threadIdx.x+i/2];
12          __syncthreads();
13      }
14      if (threadIdx.x == 0) atomicAdd(&reduxsum, blockcount[0]);
15  }
16  main() {
17      int size, *dptr, *hptr, res;
18      scanf("%d", &size);
19      dptr = cudaMalloc(sizeof(int) * size);
20      hptr = malloc(sizeof(int) * size);
21      for(int i = 0; i < size; i++) ptr[i] = rand() % 100;
22      cudaMemcpy(dptr, hptr, sizeof(int) * size, cudaMemcpyHostToDevice);
23      findsum < < < (size / 1024) + 1, 1024 > > > (ptr, size);
24      cudaMemcpyFromSymbol(&res, reduxsum, sizeof(int),
        cudaMemcpyDeviceToHost);
25      printf("%d", res);
26  }
```

5.4 Graph Analytics on GPU

The performance of topology driven graph algorithms (e.g., BFS and SSSP) varies
considerably for different graphs with the same numbers of vertices and edges.
The properties of a graph object which affect the performance of topology-driven
algorithms are *diameter*, variance in *degree* distribution, and the number of active
elements (*edges* or *vertices*). The dependence on the above properties in GPUs is
due to warp based execution, and massively parallel architecture of the GPU. Graph
algorithms have irregular access patterns and if care is not taken, this enforces usage
of atomic operations for most of the graph algorithms, which can be expensive.

5.4.1 Topology Driven Algorithms

Algorithm 5.5: Parallel Vertex-Based BFS Computation

```
1  Boolean changed; int level;
2  BFS (Vertex v, Graph G {
3  |    foreach( Vertex t In v.outnbrs ){
4  |    |    if( t.dist > v.dist + 1 ){
5  |    |    |  t.dist = v.dist + 1; changed = True;
6  |    |    }
7  |    }
8  }
9  Call_BFS(Vertex src, Graph G {
10 |    foreach( Vertex t In G.V ) in parallel {
11 |    |  t.dist = ∞;
12 |    }
13 |    src.dist = 0; level = 0; changed = True;
14 |    while( changed ){
15 |    |    changed = False;
16 |    |    foreach( Vertex t In G.V ) in parallel {
17 |    |    |    if( t.dist == level ){
18 |    |    |    |  BFS(t, G);
19 |    |    |    }
20 |    |    }
21 |    |    level++;
22 |    }
23 }
```

Atomic operations can be avoided even in cases where two or more threads write to the memory location simultaneously, provided all the threads write the same value [32]. The level based BFS Algorithm is a candidate for atomic-free implementation (see Algorithm 5.5[4]). Each vertex is typically processed by one thread (see Line 15) which sequentially examines its neighbours (see Line 3). The function BFS is called only on vertices with their *dist* value equal to the value of variable *level* (see Line 17). To begin with only the source vertex has its *dist* (distance from source, measured by the number of edges) set to zero and all others set to infinity. Variable *level* is also set to zero. Unexamined vertices get their *dist* values updated when their level is reached (see Line 3). The global variable *changed* is set to one if *dist* value is updated in one or more vertices (see Lines 3). The computation reaches a fixed point once BFS distance values of all the vertices reachable from the source vertex are computed. In next invocation, no thread (vertex) will set the *changed* variable to one, marking termination.

---

[4]A CUDA program equivalent to this algorithm must handle all the details of storing and accessing the graph, and other aspects. These are deferred until the chapter on compilation of Falcon.

It is possible that several threads (vertices) launched with the same value of *level* try to update the *dist* value of a vertex simultaneously. However, since all of them write the same value, correctness is not sacrificed. Therefore, the algorithm does not require any atomic operations (hardware guarantees atomicity of a single write for primitive data types).

BFS computation can create warp divergence, due to the fact that the number of active vertices is small during the initial and the final stages (only the source vertex is active at the beginning). However, all the threads would be launched with many having nothing to do. This happens due to the conditional statement in Line 15. Those threads whose vertices satisfy the condition will execute the BFS call statement and others will do nothing. The number of threads (vertices) which satisfy the condition in the same warp and the number of neighbours for those threads (vertices) in the BFS computation are dependent purely on the topology of the input graph.

Algorithm 5.6: Parallel Edge Based SSSP Computation

```
 1  SSSP(Graph G) {
 2      foreach( Point p In G ) in parallel {
 3          p.dist = ∞;
 4      }
 5      src.dist = 0; changed = True;
 6      while( changed ){
 7          changed = False;
 8          foreach( edge p → t In G ) in parallel {
 9              if( p.dist < t.dist ){
10                  currdist = t.dist;
11                  MIN(t.dist, p.dist + G.getWeight(p, t));
12                  if( t.dist < currdist ){
13                      changed = True;
14                  }
15              }
16          }
17      }
18  }
```

The pseudo-code for the edge based SSSP computation from Algorithm 3.7 is reproduced in a slightly modified form in Algorithm 5.6. Edge based computation results in *fine-grained* parallelism where computation per thread is very small. In vertex based SSSP computation shown in Algorithm 3.5, each vertex is processed by a thread, which in turn, sequentially processes all its outgoing neighbours. This results in *coarse-grained* parallelism with more work per thread. However, in general, edge based processing results in less warp divergence than vertex based processing.

The performance of topology-driven algorithms on a GPU is dependent on the *diameter* and *degree distribution* of the graph object. The larger the diameter, the lesser the parallelism in each invocation of the kernel. The above phenomenon is observed in road-network graphs, which have very high diameters, very low degree, and very low variance in degree. Such graphs need worklist based processing, but this type of processing is inefficient on GPUs (see Sect. 5.4.2). Social network graphs have low diameters, but they have high variance in degree distribution which leads to warp divergence in vertex based processing. The variance in degree distribution makes edge based processing more efficient as it reduces warp divergence.

Algorithm 5.6 requires the atomic operation MIN (see Line 10) as two different threads $t1$ and $t2$ may try to reduce the *dist* value stored in the same memory location using two edges $t1 \rightarrow t3$ and $t2 \rightarrow t3$ with different edge weights w_1 and w_2 respectively. The values written to $t3.dist$ by the two threads could be different for different edge weights. The *atomic* MIN operation is required to ensure semantic correctness. The *atomic* operation serializes the memory updates at runtime and this leads to poor performance.

5.4.2 Other Elementary Graph Algorithms

Warp divergence is present in vertex based algorithms, which iterate over vertices and their neighbours. This can be eliminated by iterating over edges. The performance effects will be visible when the variation in degree distribution is very high. Other elementary algorithms such as *connected components*, *minimum spanning tree* etc., do not follow a topology driven processing as in algorithms such as BFS and SSSP. *Diameter* of a graph does not affect the performance of these benchmarks. These algorithms involve graph contraction and can be implemented using the disjoint set union-find data structure.

Worklist based implementation of algorithms does not always perform well on GPUs. The poor performance in such cases is due to atomic addition of elements to the worklist and lack of parallelism present in such implementations to use massively parallel GPU device. Δ-stepping (bucket based) implementation of elementary algorithms is very efficient on multi-core CPUs but not on GPUS. The parallelism is not enough to utilize more than thousand cores available in GPUs, but good enough for less than hundred cores of a multi-core CPU.

Chapter 6
Dynamic Graph Algorithms

Dynamic graph algorithms compute the graph properties from the previous set of values. Typical operations in dynamic graph algorithms are insertion and deletion of edges and vertices, and the query for property values relevant to the algorithm. The efficiency of a dynamic algorithm depends on the data structure used to implement it. This chapter provides a glimpse into this exciting area in graph analytics.

6.1 Introduction

Real world graphs change their topology over time. Social network graphs get modified with addition and deletion of edges and vertices. A typical example is the twitter network graph,[1] with users as vertices and *"following"* relationship between users as edges. Road network graphs where vertices represent junctions and edges represent roads, get updated by the addition and deletion of roads. The change in topology in road network graphs happens over a longer period of time when compared to social network graphs. The edge weights of road network graphs which represent expected travelling time from one junction (source vertex) to another junction (destination vertex) change frequently based on traffic congestion. Traffic congestion results in change in shortest paths multiple times over a day. The shortest distance route changes over a long period of time when a road gets added or deleted. The deletion of edge happens in situations such as a bridge getting damaged. The *movie-actor* or *author-publication* bipartite graphs change by the addition of vertices and edges. Algorithms such as Delaunay Mesh Refinement (DMR) and Delaunay Triangulation (DT) used in computational geometry are also dynamic where the mesh of triangles is refined by the addition of points, and addition/deletion of edges.

---

[1] https://snap.stanford.edu/data/twitter-2010.html.

© Springer Nature Switzerland AG 2020
U. Cheramangalath et al., *Distributed Graph Analytics*,
https://doi.org/10.1007/978-3-030-41886-1_6

Developing efficient graph analytics for dynamic graphs is challenging. Not many frameworks exist which provides API to program dynamic graph analytics. The Falcon DSL [26] supports dynamic graph analytics by providing API functions on Graph datatype to change the topology of the graph at runtime.

Dynamic graph algorithms are called *incremental* if they operate on dynamic graphs in which graph elements are added, but not deleted. Some algorithms such as the shortest paths computation are incremental when edge-weights *increase*. Graph algorithms for *movie-actor* or *author-publication* graphs are naturally incremental. *Incremental* dynamic graph algorithms result in the shortest path distance, MST cost, number of connected components etc., to *increase* or remain the same after each increase in weight. Deletion of an edge e with weight w can also be considered as increasing the weight of edge e from w to ∞, for algorithms in which a minimum on weights is computed (for example, SSSP, MST, etc.).

Decremental graph algorithms deal with deletion of graph elements or a *decrease in weights* of edges. Decremental graph algorithms result in the shortest path distance, MST cost, number of connected components etc., to *decrease* or remain the same after each decrease in weight. Addition of an edge e with weight w can be considered as reducing the weight of edge e from ∞ to w for algorithms in which a minimum on weights is computed (for example, SSSP, MST, etc.).

Algorithms operating on graphs in which vertices and edges are both added and deleted are *fully dynamic* graph algorithms. In these algorithms, increase and decrease of edge weights is also handled. The properties associated with the vertices and edges can increase or decrease based on how the graph topology is affected. Dynamic graph analytics involves recomputing new properties from the current values rather than computing them from scratch. The recomputation is done at periodic intervals or on demand.

Figure 6.1 shows how the shortest distance path from source vertex v_0 to target vertex v_4 on the input graph in Fig. 6.1b is modified on the deletion of the edge $v_0 \rightarrow v_1$ (see Fig. 6.1a) and on the addition of the edge $v_3 \rightarrow v_4$ (see Fig. 6.1c). The edges belonging to the shortest path are shown in bold font.

Figure 6.2 shows how the two connected components of an undirected input graph in Fig. 6.2b change on deletion of the edge e_1 (see Fig. 6.2a, which shows

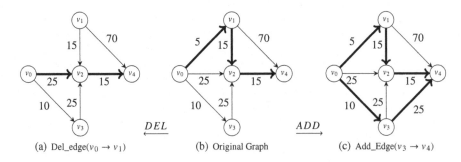

(a) Del_edge($v_0 \rightarrow v_1$)   (b) Original Graph   (c) Add_Edge($v_3 \rightarrow v_4$)

Fig. 6.1 Shortest distance path ($v_0 \rightarrow v_4$): Del_ Edge(), original graph, and Add_Edge()

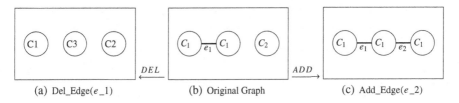

(a) Del_Edge(e_1)   (b) Original Graph   (c) Add_Edge(e_2)

Fig. 6.2 Connected components: Del_Edge(), original graph, Add_Edge()

three connected components) and on addition of the edge e_2 (see Fig. 6.2c, which shows a single connected component).

6.2   Dynamic Algorithms for Elementary Graph Problems

Typical operations happening on dynamic graphs are *insertion, deletion, update,* and *query. Insertion* and *deletion* can be for edges or vertices or both. The *update* operation updates the properties of the graph object and *query* operation fetches the property values. These operations ought to be efficient. The property values are computed from the current values on change in topology of the graph. The computational complexity of dynamic graph algorithms is explored in [150, 151].

6.3   Dynamic Shortest Path Computation

There are optimizations possible with deletion and addition of edges for the Single Source Shortest Path (SSSP) problem. Deletion of an edge e is considered here as increasing its weight to ∞. Addition of an edge e with weight w is considered here as reducing the weight of e from ∞ to w (provided there is no edge with same source and destination vertex currently in the graph object with weight less than w). Dynamic *single destination shortest path* (SDSP) algorithms using priority queues which compute the shortest path from current values have been reported in literature [152]. SSSP computation on a graph G can be simply performed by computing SDSP on the reverse or transpose graph of G (all edges reversed), with the source marked as the destination. It is also possible to make simple modifications to the SDSP algorithm in Algorithm 6.1 to directly compute SSSP.

The incremental and decremental SDSP algorithms store properties related to shortest path computation. Let T $\subseteq$ G be an acyclic graph which contains all the vertices of G and all the edges belonging to all the single destination shortest paths. The graph becomes an acyclic graph when multiple shortest paths exist for the same vertex; otherwise it is a tree. The algorithms use the data structures mentioned below.

The first three data structures are assumed to be maintained dynamically and should be available when updates need to be applied:

1. *spe* is a boolean array of size $|E|$: $spe[i] = 1$ if $edge_i \in$ T, otherwise 0.
2. *spd* is an array of size $|V|$: $spd[i] =$ shortest path distance from v_i to the destination vertex.
3. *vcnt* is an array of size $|V|$: $vcnt[i] =$ count of edges $e : v_i \to v_j$ which are part of T, i.e., $spe[e] = 1$.
4. Q is a set of vertices.
5. *pqueue* is a priority queue implemented using a heap.

6.3.1   Incremental Dynamic SDSP Computation

The pseudo code for the *incremental dynamic SDSP computation* is shown in Algorithm 6.1 where *edge weight is increased* for a *single edge*. This can handle deletion of an edge as well. The function takes as arguments, the edge $e : v1 \to v2$, whose weight is increased, edges belonging to T (*spe*), the distance value for each vertex to the destination (*spd*), the number of edges belonging to T from each vertex (*vcnt*), and the graph object (*G*). The Heap object used for the SDSP (Single Destination Shortest Path) recomputation is *pqueue*, which stores vertices of the graph object. The algorithm simply returns if the edge e does not belong to T (see Line 4). The increase in the weight does not affect the shortest path in this case. If the edge e belongs to T, value *vcnt[v1]* is decreased by one. The shortest path values of the vertices need to be recomputed if and only if there are no other outgoing edges from $v1$ which belong to T, (that is if $vcnt[v1]$ becomes zero). The function returns immediately, if $vcnt[v1]$ is greater than zero (see Line 6).

The loop at Line 8 identifies and accumulates all the vertices affected by the edge whose weight is increased, and places these vertices in the set Q. These are the vertices connected to the source of the deleted edge directly or indirectly by the shortest path tree edges. It also unmarks all the edges in the shortest path tree that are connected to the vertices in Q. The next loop at Line 19 tries to compute tentative values for the shortest paths of the vertices in Q, and places them in the Heap object *pqueue*. Then the while-loop at Line 25 uses the Heap object *pqueue* to consider one vertex at a time in the order of increasing distance from the destination and recomputes the distances. It also recomputes the shortest path tree in the affected region simultaneously.

The vertex *v1* is added to the set Q. The *spd* value of all the vertices added to the queue is made ∞ (Line 9). The shortest path from vertex *v1* to the destination vertex needs to be recomputed now. The other affected vertices are now added to Q (see Lines 8–18). The shortened distance needs to be recomputed for each vertex t such that $t \to v \in T, v \in Q$, and $vcnt[t] == 1$ (as discussed for vertex v1). If $vcnt[t] - 1$ is zero, t is considered as affected by the increase in weight of the edge $v1 \to v2$, and is added to Q.

Algorithm 6.1: Incremental SDSP Algorithm

```
1   Incr_SDSP(Edge e, SP_edges spe, SP_dist spd, V_cnt vcnt, Graph G) {
2       Vertex v1 = e.src, v2 = e.dst;
3       Heap < Vertex > pqueue;
4       if (spe[e] == 0)return;
5       vcnt[v1] = vcnt[v1] - 1;
6       if (vcnt[v1] > 0) return;
7       Set Q = {v1};
8       foreach( v In Q ){
9           spd[v] = ∞;
10          foreach( t In v.innbrs ){
11              Edge e = G.getEdge(t, v);
12              if ( (spe[e] == 1 ){
13                  spe[e] = 0;
14                  vcnt[t] = vcnt[t] - 1;
15                  if(vcnt[t] == 0) Q = Q∪{t};
16              }
17          }
18      }
19      foreach( v In Q ){
20          foreach( t In v.outnbrs ){
21              if( spd[v] > spd[t] + Graph.getWeight(v, t)) spd[v] = spd[t] +
                    Graph.getWeight(v, t);
22          }
23          if(spd[v] ≠ ∞) pqueue.add(v, spd[v]);
24      }
25      while( pqueue.size > 0 ){
26          Vertex v = pqueue.delMin();
27          foreach( t In v.innbrs ){
28              if ( spd[t] > spd[v] + Graph.getWeight(t, v) ){
29                  spd[t] = spd[v] + Graph.getWeight(t, v);
30                  pqueue.adjust(t, spd[t]);
31              }
32          }
33          foreach( t In v.outnbrs ){
34              if ( spd[v] == Graph.getWeight(v, t) + spd[t] ){
35                  spe[v → t] = 1;
36                  vcnt[t]++;
37              }
38          }
39      }
40  }
```

The elements in Q are then processed (see Lines 19–24). For an element $u \in Q$, all the outgoing edges $u \rightarrow v$ are considered and if $spd[v] + weight(u \rightarrow v)$ is greater than $spd[u]$, $spd[u]$ is modified to $spd[v] + weight(u \rightarrow v)$. If the distance value is reduced from ∞ for $u \in Q$, it is added to the heap *pqueue*.

The elements in *pqueue* are then processed (see Lines 25–39). The vertex v with minimum distance value is deleted from *pqueue*. Then all the vertices t such that the edge $t \rightarrow v \in G.E$ are considered and the distance value of t is reduced if its current value is greater than $dist[v] + weight(t \rightarrow v)$. If the value is decreased, then the vertex t is added to *pqueue* using the *adjust* function. The *adjust* function decreases the value of the vertex t if it is already in *pqueue*. Otherwise, it adds t to *pqueue*. T and the number of outgoing edges in T for each vertex are then updated in a loop (see Lines 33–38). The process continues until *pqueue* becomes empty.

6.3.1.1 Example for Incremental SDSP Computation

Figure 6.3 shows an input graph in Fig. 6.3a and the modified graph in Fig. 6.3b on deletion of the edge $v_2 \rightarrow v_0$. To compute single destination shortest path (SDSP), the destination vertex is taken as v_0 for the example. The edges which are part of the SDSP tree T are shown in *bold face* in both the subfigures.

The edge $v_2 \rightarrow v_0$ is a part of T and the vertex v_2 has V_{cnt} value of one. So the vertex v_2 is added to the set Q. All the incoming edges to v_2 are now processed. The vertices v_4 and v_3 are also added to Q, as the V_{cnt} value becomes zero for both the vertices. Vertex v_1 is not considered as the edge $v_1 \rightarrow v_2$ is not a part of T. The set Q now contains the vertices v_2, v_4 and v_3, but they need not be processed in that order (Q is a set). The distance field of all the elements in Q is set to ∞.

The vertices in Q are now considered in any arbitrary order, say, in the order v_3, v_4 and v_2 (this order illustrates the example better). The outgoing edges of v_3 are considered first. Edge $v_3 \rightarrow v_2$ contributes nothing since v_2 is ∞. Edge $v_3 \rightarrow v_0$ makes the tentative value of spd[v_3] as 45, since $infty > spd[v_0] + 45 = 0 +$

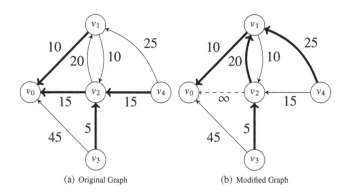

(a) Original Graph   (b) Modified Graph

Fig. 6.3 Decremental single destination shortest path

$45 = 45$. Thus, $(v_3, 45)$ is added to the heap *pqueue*. Now vertex v_4 is considered. As before, the edge $v_4 \rightarrow v_2$ yields nothing. the edge $v_4 \rightarrow v_1$ makes spd[v_4]as spd[v_1] $+ 25 = 10 + 25 = 35$, tentatively. Therefore, $(v_4, 35)$ is now added to the heap, *pqueue*. Vertex v_2 is processed next. Edge $v_2 \rightarrow v_1$ sets spd[v_2] as spd[v_1] $+ 20 = 10 + 20 = 30$, tentatively, and $(v_2, 30)$ is also added to the heap, *pqueue*.

The elements of the heap *pqueue* processed in increasing order of the distance component (spd[]) of its elements. The first element to be processed is $(v_2, 30)$. Its in_neighbour v_3 (with spd[v_3] as 45) gets spd[v_3] reduced to 35 due to the edge $V_4 \rightarrow V_2$, because $spd[V_2] + 5 = 30 + 5 = 35 > 45$. The entry $(v_3, 45)$ in the heap gets modified to $(v_3, 35)$. The other two in_neighbours v_1 and v_4 are not affected. While processing out_neighbours, the edge $v_2 \rightarrow v_1$ is set as a tree edge.

The element $(v_4, 35)$ is removed from *pqueue* and processed next. There are no in_neighbours to be processed. The out_neighbour loop marks the edge $v_4 \rightarrow v_1$ as a tree edge. Now the last element $(v_3, 35)$ is removed from the heap and processed. Again, there are no in_neighbours to be processed. The out_neighbour loop marks the edge $v_3 \rightarrow v_2$ as a tree edge. The algorithm now terminates.

Table 6.1 shows the initial and final values of V_{cnt} and SP_{dist} for all the vertices.

The algorithm is efficient when it accepts all the edges in a single batch. The set Q can be populated with the source vertices of all the deleted edges right in the first step. The rest of the algorithm remains the same.

Modifying the incremental SDSP algorithm to compute SSSP is simple. The destination vertex of the deleted edge is added to Q, instead of the source vertex (v_1 is replaced by v_2 in Algorithm 6.1). The out_neighbours are processed first, followed by in_neighbours in the two loops at Lines 8 and 19, and also in the two loops at Lines 27 and 33.

As an example of computing SSSP incrementally, Fig. 6.4 shows graphs with shortest paths before and after deleting the edge $v_0 \rightarrow v_2$. The edges belonging to T are shown in bold face. The source vertex is taken as v_0. Table 6.2 shows the initial and final values of V_{cnt} and SP_{dist} for all the vertices.

Table 6.1 SP_{dist} and V_{Cnt} values (initial and final) on deletion of the edge $v_2 \rightarrow v_0$

Vertices	v0	v1	v2	v3	v4
V_{cnt} (initial)	0	1	1	1	1
SP_{dist} (initial)	0	10	15	20	30
V'_{cnt} (final)	0	1	1	1	1
SP'_{dist} (final)	0	10	30	35	35

Table 6.2 SP_{dist} and V_{Cnt} values initial and final on deletion of the edge $v_0 \rightarrow v_2$ (see Fig. 6.4)

Vertices	v0	v1	v2	v3	v4
V_{cnt} (initial)	2	0	2	0	0
SP_{dist} (initial)	0	10	15	20	30
V'_{cnt} (final)	1	2	1	0	0
SP'_{dist} (final)	0	10	30	35	35

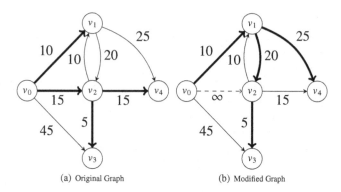

(a) Original Graph   (b) Modified Graph

Fig. 6.4 Incremental single source shortest path

6.3.2 Decremental Dynamic SDSP Computation

The decremental version of SDSP allows for insertion of a new edge with a weight
w, which is treated as reducing the weight of the edge from ∞ (non-existent edge) to
w. The pseudo-code for the decremental SDSP algorithm is shown in Algorithm 6.2.
If the new edge $v_1 \rightarrow v_2$ does not change the distance of vertex v_1 to the destination
(see Line 4), the processing halts. If the new edge creates an alternate shortest path
from v_1 to destination, it is marked as a shortest path edge and the path counter
vcnt[v_1] is incremented (see Line 7). Otherwise, distance of v_1 is updated and the
heap *pqueue* in initialized with v_1.

The While-loop at Line 12 removes a minimum distance vertex at a time from the
heap *pqueue* and processes its out_neighbours and then its in_neighbours. The for-
loop at Line 15 processing out_neighbours of vertex v, either adds an alternative
shortest path passing through v, the edge $v \rightarrow t$, and t, or it ignores the edge $v \rightarrow t$.
The for-loop at Line 24 processing in_neighbours of vertex v, updates the shortest
path counter of t, if passing through t, the edge $t \rightarrow v$, and v, is cheaper than taking
an alternative path through t bypassing v, and adds t to the heap, *pqueue*. Otherwise,
the edge $t \rightarrow v$ is marked as part of the shortest path, if the shortest path passing
through t, the edge $t \rightarrow v$, and v is an alternative shortest path (see Line 25).

6.3.2.1 Example for Decremental SDSP Computation

As an example, consider adding the edge $v_2 \rightarrow v_0$ with weight 15, back to the graph
in Fig. 6.3b. Vertex v_2 gets added to the heap, *pqueue*, along with a distance of 15.
While processing the out_neighbour v_0 of v_2, the edge $v_2 \rightarrow v_0$ gets marked as a
shortest path edge. There is no effect on v_1. While processing its in_neighbour v_3,
the shortest path from v_3 reduces and it gets added to the heap, *pqueue*, along with
a distance of 20. Similarly, v_4 also gets added to the heap along with a distance
of 30. In the next two iterations of the while-loop, the two edges, $v_3 \rightarrow v_2$ and

---

Algorithm 6.2: Decremental SDSP Algorithm

```
 1  Decr_SDSP(Edge e, SP_edges spe, SP_dist spd, V_cnt vcnt, Graph G) {
 2      Vertex v1 = e.src,v2 = e.dst;
 3      Heap <Vertex> pqueue;
 4      if( spd[v1] < spd[v2] + Graph.getweight(v1, v2) ){
 5          return;
 6      }
 7      if( spd[v1] == spd[v2] + Graph.getweight(v1, v2) ){
 8          spe[e] = 1; vcnt[v1]++; return;
 9      }
10      spd[v1] = spd[v2] + Graph.getweight(v1, v2);
11      pqueue.add(v1, spd[v1]);
12      while( pqueue.size > 0 ){
13          Vertex v = pqueue.del();
14          vcnt[v] = 0;
15          foreach( t In v.outnbrs ){
16              if( spd[v] == spd[t] + Graph.getweight(v, t) ){
17                  spe[v→t] = 1;
18                  vcnt[v]++;
19              }
20              else{
21                  spe[v→ t] = 0;
22              }
23          }
24          foreach( t In v.innbrs ){
25              if( spd[t] > Graph.getWeight(t, v) + spd[v] ){
26                  spd[t] = spd[v] + Graph.getweight(t, v);
27                  pqueue.adjust(t, spd[t]);
28              }
29              }else if { spe[t→v] == 0 && spd[t] == Graph.getWeight(t, v) + spd[v] ){
30                  spe[t→v] = 1;
31                  vcnt[t]++;
32              }
33          }
34      }
35  }
```

$v_4 \rightarrow v_2$, get marked as being on the shortest paths. The algorithm terminates thereafter, yielding the graph in Fig. 6.3a. It may be noted that adding the edge $v_2 \rightarrow v_0$ with a weight of 20, changes the shortest path from v_3, but simply adds an alternative shortest path from v_4. A weight of 25 adds an alternate shortest path from v_3, but does not change the shortest path from v_4. However, a weight of 30 adds an alternative shortest path from v_2 but does not change the other shortest paths. A weight more than 30 does not change any shortest path.

A decremental version of SSSP can be designed easily from the corresponding SDSP algorithm on the same lines as described in Sect. 6.3.1.1.

6.4   Computational Geometry Algorithms

Algorithms from the domain of computational geometry which operate on meshes can be viewed as graph algorithms where multiple edges make a mesh. Examples of such meshes are triangular grids and rectangular grids. Algorithms such as Delaunay Mesh Refinement and Delaunay triangulation are dynamic where new vertices and edges are added when triangles are deleted and new triangles are added. Delaunay mesh refinement algorithms [153] are meant for construction of meshes of triangles or tetrahedra. They are suitable in several applications such as interpolation, rendering, terrain databases, geographic information systems, and importantly, the solution of partial differential equations by the finite element method.

6.4.1   Delaunay Triangulation (DT)

A triangulation (in geometry) is a subdivision of a planar object into triangles. A 2D Delaunay mesh is a triangulation of a set of points that satisfy the following property: no other point from the mesh must be contained in the circumcircle of any triangle. Algorithm 6.3 shows an outline of Delaunay triangulation for a set of points. First of all, a super triangle that is large enough to hold all the points is created. This super triangle need not be precise; it can be much larger than what is needed. Mesh M contains all the triangles after a proper triangulation. A set of all the triangles in M which become *bad* due to insertion of a point from the list of points to be triangulated, is first collected. Then, all the edges lining the cavity of bad triangles are collected, and all the bad triangles are removed from M. The cavity is now re-triangulated, and the original mesh is updated by adding the new triangles created by re-triangulation. The order in which the bad triangles are processed is irrelevant (the final mesh may differ based on the order, but the geometric criteria can still be satisfied). The above procedure is repeated by inserting one point at a time from the list of points to be triangulated.

6.4.2   Delaunay Mesh Refinement (DMR)

These algorithms operate by maintaining a Delaunay triangulation which is refined by inserting additional vertices until the mesh meets requirements on element quality and size. A refined Delaunay mesh meets the additional quality constraint (e.g., no triangle has an angle less than 30°). Algorithm 6.4 takes a Delaunay mesh as input. This mesh may contain triangles that do not meet the quality constraints (called bad triangles). Iterative re-triangulation of the affected portions of the mesh produces a refined mesh, with no bad triangles. This procedure is similar to the

---

Algorithm 6.3: Delaunay Triangulation Algorithm

```
 1  DT(ListofPoints) {
 2        // ListofPoints is a set of points to be triangulated
 3        Mesh M = super_triangle;
 4        // Must be large enough to accommodate the whole ListofPoints
 5        Collection bad_Triangles = empty set;
 6        foreach( Point p In ListofPoints ){
 7              // add points one at a time to the triangulation
 8              foreach( Triangle t In M ){
 9                    // find all the triangles that p invalidates
10                    if( Point p is inside circumcircle of Triangle t ){
11                          add t to bad_Triangles;
12                    }
13              }
14              Collection cavity := empty set;
15              foreach( Triangle t In bad_Triangles ){
16                    // find the boundary of cavity
17                    foreach( edge e In t ){
18                          if( edge e is not shared with any other triangle in bad_Triangles ){
19                                add e to cavity;
20                          }
21                    }
22              }
23              foreach( Triangle t In bad_Triangles ){
24                    // remove it from the data structure
25                    remove t from M;
26              }
27              foreach( edge e In cavity ){
28                    // re-triangulate cavity and update M
29                    Triangle new_Tri = form a triangle from vertices of e to p;
30                    add new_Tri to M;
31              }
32        }
33        foreach( Triangle t In M ){
34              // Inserted points and triangulated. Clean up
35              if( Triangle t contains a vertex from original super_triangle ){
36                    remove t from M;
37              }
38        }
39        return M;
40  }
```

Fig. 6.5 Delaunay
triangulated mesh

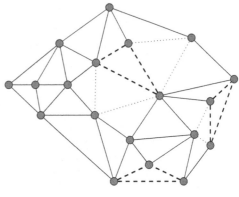

Fig. 6.6 Delaunay refined
mesh

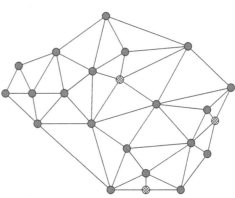

triangulation procedure described in Algorithm 6.3. Figure 6.5 shows a delaunay
triangulated mesh which is an input to the DMR algorithm. The bad triangles are
shown using dashed-lines and the triangles belonging to the cavity near the bad
triangles are shown using dotted-lines. The DMR algorithm refines such a mesh.
The mesh generated by the DMR algorithm is shown in Fig. 6.6, which satisfies the
condition that all the angles in all the triangles are greater than 30°. Newly inserted
points in Fig. 6.6 are filled with cross pattern.

6.4.3 Parallel Delaunay Mesh Refinement (DMR)

The pseudo-code for parallel computation of DMR is shown in Algorithm 6.5. The
algorithm first computes the set of all bad triangles in the input delaunay triangulated
mesh (Line 5). The set of all bad triangles is stored in the dynamic collection
(varying size) *bad_triangles*. The algorithm re-triangulates all the elements in
the Collection object *bad_triangles* in parallel (Lines 7–15). The cavity for
each bad triangle contains a collection of its surrounding triangles which needs

---

Algorithm 6.4: DMR Algorithm

```
 1  DMR() {
 2      Mesh M;
 3      read_mesh(M);
 4      Collection bad_triangles;
 5      bad_triangles = getBadTriangles(M);
 6      foreach( Triangle t In bad_triangles ){
 7          Cavity cav = new Cavity(tria);//Cavity is a Collection type
 8          expand(cav);
 9          retriangulate(cav);
10          updateMesh(cav, M);
11          add(bad_triangles, getBadTriangles(cav));
12      }
13  }
```

---

to be re-triangulated. During parallel computation, two bad triangles *btr1*, *btr2* ∈ *bad_triangles* may belong two different cavities, but with one or more triangles in common. Concurrent modification of the same triangle by two or more threads may lead to incorrect computation. Therefore, each thread should acquire a lock on all the triangles in its cavity. Algorithm 6.5 shows such a locking procedure using a `single` statement which tries to lock all the elements in the `Collection` object *cav* local to each thread (Lines 10–14). If two or more threads have common triangles among their cavities the `single` statement makes sure that only one thread succeeds and priority is given to the thread with the lowest thread id. The other threads are processed in the next iteration. The algorithm terminates when all the bad triangles are processed which brings down the number (size) of *bad_triangles* to zero (Line 16). The re-triangulation of a cavity can add new elements to the `Collection` object *bad_triangles* (Line 13).

The parallel DMR algorithm, which uses a lock on the collection of items by each thread requires a barrier for the entire code. OpenMP supports such a barrier for a multi-core CPU. Such a global barrier is not supported in CUDA, and must be implemented in software. The CUDA kernel should be called with the number of thread blocks equal to the number of SMs in the GPU. Each thread in the GPU may then process one or more elements based on the number (size) of bad triangles. A global barrier for CUDA kernel was discussed in Algorithm 5.2 (Chap. 5) and a similar feature is required for parallel implementation of DMR in GPU [26].

6.5 Challenges in Implementing Dynamic Algorithms

The major challenges in efficient implementation of dynamic graph algorithms are now highlighted.

Algorithm 6.5: Parallel DMR Algorithm

```
 1  Parallel_DMR() {
 2      Mesh M;
 3      read_mesh(M);
 4      Collection bad_triangles;
 5      bad_triangles = getBadTriangles(M);
 6      while( True ){
 7          foreach( Triangle t In bad_triangles ) in parallel {
 8              Cavity cav = new Cavity(tria);//Cavity is a Collection type
 9              expand(cav);
10              single( cav ){
11                  retriangulate(cav);
12                  updateMesh(cav, M);
13                  add(bad_triangles, getBadTriangles(cav));
14              }
15          }
16          if(bad_triangles.size == 0) break;
17      }
18  }
```

6.5.1 Dynamic Memory Management

The change in graph topology due to addition and deletion of vertices and edges demands efficient utilization of memory. Deletion of edges and vertices demands efficient garbage collection, which is programmer's responsibility in unmanaged languages such as C and C++. Addition of edges requires the graph object to increase its size at runtime. This requires reallocation of the graph memory so that more edges and vertices can be accommodated at runtime. Compilers such as *gcc* and *nvcc* provide library functions for reallocation of memory. Efficient utilization is possible only when deleted memory locations (occupied by deleted edges and vertices) are freed and some compaction is performed so that garbage memory is minimal. This is crucial when the data sizes are huge, but requires more effort from the programmer.

6.5.2 Parallel Computation

Most elementary dynamic graph algorithms require different data structures such as Queue, *Heap*, *Forest of trees*, etc. Efficient parallel implementation of these data structures is very challenging. The performance benefit of parallel implementation of elementary dynamic graph algorithms using these data structures is still debatable. There are not many parallel implementations of dynamic algorithms available in literature. Algorithms such as DMR and DT require a lock for the entire parallel region (*global barrier*). Such a feature is missing in GPU devices, and needs to be implemented in software, as discussed in Algorithm 5.2. Such global barriers on GPUs are inefficient.

However, it must be said in defense of dynamic algorithms that even though these are sequential, these may be faster than non-dynamic parallel algorithms for the same tasks. This is because dynamic algorithms operate only on a small part of the graph whereas non-dynamic algorithms operate on the whole graph. Therefore, depending upon the situation, one may choose not to parallelize dynamic algorithms.

Chapter 7
`Falcon`: A Domain Specific Language for Graph Analytics

The domain-specific language `Falcon` is presented in this chapter. The data types and statements of `Falcon` that support easy programming of graph analytics applications are described. To drive home the point that `Falcon` programs can be very efficient, code generation mechanisms used in the `Falcon` compiler are delineated with examples.

7.1 Introduction

Graph analytic frameworks provide abstractions which enforce the algorithmic logic to be written inside the API functions specific to each framework. The programmer may still be required to handle dynamic memory allocation and thread management. Domain Specific Languages (DSLs) provide a higher level of abstraction with special constructs and data types specific to the domain. A DSL program is closer to the pseudo-code of an algorithm. This eases programming and enhances productivity while producing code for applications specific to the domain. Of course, this benefit is often at the cost of generality, and one may not be able to implement algorithms outside that domain in the corresponding DSL. `SQL` for database applications and `HTML` for web programming are examples of DSLs. This chapter discusses DSLs for graph analytics with emphasis on the `Falcon` graph DSL [27].

The novel features of the `Falcon` DSL and its compiler are listed below:

- Support for heterogeneous hardware: `CPU` and `GPU`
- Support for programming graph analytics on distributed heterogeneous systems
- Support for dynamic graph algorithms
- Support to represent and process meshes as graphs

© Springer Nature Switzerland AG 2020

U. Cheramangalath et al., *Distributed Graph Analytics*,

https://doi.org/10.1007/978-3-030-41886-1_7

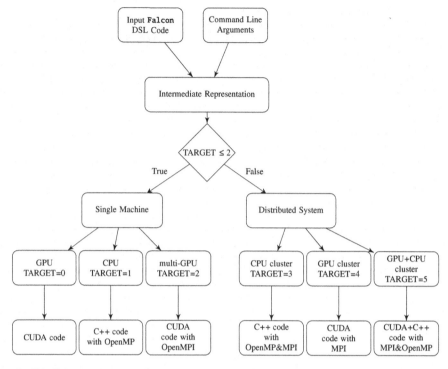

Fig. 7.1 Falcon: code generation overview

- Incorporates an efficient code generator which generates programs in native language for different hardware from a single DSL code (C++/CUDA/OpenCL etc.).
- Enables a programmer to concentrate only on the algorithm logic. Thread and memory management are handled automatically by the Falcon compiler (see Fig. 7.1).

7.2 Overview

Falcon extends the C programming language with additional data types, properties for data types, and constructs for parallel execution and synchronization. A programmer inputs a Falcon program (DSL code) and one or more command-line arguments to the Falcon compiler. The command-line arguments include *TARGET* architecture, type of the DSL program (static or dynamic), dimensions of vertices in the graph objects (supports up to three dimensional coordinates), etc. Falcon programs are explicitly parallel. The Falcon compiler also supports parallel reduction operations. An important implication of extending a general-purpose language is that the programmer can continue to write general programs with the new DSL as well.

7.2.1 SSSP Program in `Falcon`

An SSSP program in `Falcon` is shown in Algorithm 7.1. The `Graph` object *graph* is added with a *vertex* property *dist* of type int in Line 8. The *addPointProperty()* function is used to add a new property to each *vertex* in the graph object. The graph object is read in Line 9. The *dist* property of each vertex is initialized to a sufficiently large value in Line 11 using the `foreach` parallel construct. The `foreach` parallel construct operates on the `Graph` object *graph* using the iterator *points*. It is converted to parallel code by the code generator of the `Falcon` compiler. For example, `OpenMP` and `CUDA` code are generated for multi-core `CPU` and `GPU` respectively.

Algorithm 7.1: SSSP Program in `Falcon`

```
1  int changed = 0; // Global variable
2  relaxgraph(Point p, Graph graph) {
3        foreach (t In p.outnbrs)
4        MIN(t.dist, p.dist + graph.getWeight(p, t), changed);
5  }
6  main(int argc, char *argv[]) {
7        Graph graph;
8        graph.addPointProperty(dist, int);
9        graph.read(argv[1]);
10       //make dist infinity for all points.
11       foreach (t In graph.points) t.dist = 1234567890;
12       graph.points[0].dist = 0; // source has dist 0
13       while( 1 ){
14             changed = 0;
15             foreach (t In graph.points) relaxgraph(t, graph);
16             if (changed == 0) break; // reached fixed point
17       }
18 }
```

The *dist* property of the source vertex is initialized to zero in Line 12. SSSP computation then happens in the `while` loop (Lines 13–17) until a fixed point is reached. After initializing *changed*, the *dist* property values of the vertices are reduced by calling the *relaxgraph()* function using a parallel `foreach` statement. The *relaxgraph()* function takes a `Point` object *p* and a `Graph` object *graph*, as arguments. For all the edges $p \rightarrow t \in G$, the *dist* value of the vertex *t* is reduced using the *MIN()* function, provided *dist* value of *t* is greater than the sum of *dist* value of the vertex *p* and weight of the edge $p \rightarrow t$. If the *dist* value of the vertex *t* is reduced, the variable *changed* is set to one. The Falcon compiler relies on *coarse-grained parallelism* and does not support nested parallelism (see discussion in Sect. 3.2.1). The foreach statement inside the relaxgraph function (Line 3) is converted to a sequential for loop in the generated code as it is nested under the foreach statement which calls the *relaxgraph()* function in Line 15. The *MIN()* function is converted to *atomic-min()* function of GPU or CPU based on the *target* hardware. The semantics of the MIN(arg1, arg2, changed) library function of Falcon is as follows:

atomic if (arg1 > arg2) {arg1 = arg2; changed = 1;}

If the value of the variable *changed* is still zero in Line 16, it implies that distances have stabilized and have not changed for any vertex in the last iteration. The fixed-point computation terminates in this case. Otherwise, the while loop continues for another iteration.

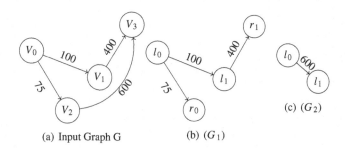

(a) Input Graph G   (b) (G_1)   (c) (G_2)

We now describe how a distributed processing is generated by the Falcon compiler with SSSP as an example. An important aspect in a distributed setup is partitioning the underlying data. Graph partitioning is required when the graph is large and computation happens on more than one device. Consider an example graph with four vertices is shown in Fig. 7.2a. The graph is partitioned into two subgraphs G_1 (Fig. 7.2b) and G_2 (Fig. 7.2c) for a distributed computation. The Falcon compiler uses *edge-cut* graph partitioning. The subgraph G_1 has V_0 and V_1 as local vertices (renamed inside G_1 as l_0 and l_1 respectively), and V_2 and V_3 as remote vertices (renamed inside G_1 as r_0 and r_1 respectively) through the edges $V_0 \rightarrow V_2$ and $V_1 \rightarrow V_3$. The subgraph G_2 has vertices V_2 and V_3 as local vertices (renamed inside G_2 as l_0 and l_1), and no remote vertices. All the outgoing edges from the local vertices are stored in the respective partitions (subgraphs). Tables 7.1 and 7.2 show the edge-list and index arrays of the subgraphs G_1 and G_2. The element *index*[*i*] stores the position of the first outgoing edge from vertex *i* in the edge-list array. The size of *index* array is ($|V| + 1$) and the value stored in (index[$|V| + 1$]) is $|E|$. The number of outgoing edges from a vertex *p* is found from the value ($index[p + 1] - index[p]$).

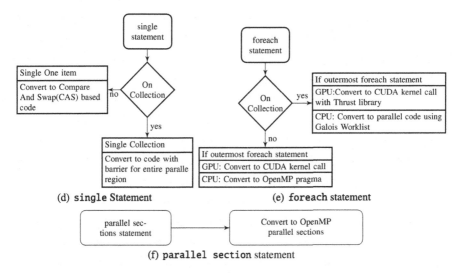

Fig. 7.2 Overview: parallel constructs code generation (single machine)

Table 7.1 Edge-list for subgraphs G_1 and G_2

Subgraph	dst	Weight	dst	Weight	dst	Weight
G_1	$V_1(l_1)$	100	$V_2(r_1)$	75	$V_3(r_2)$	400
G_2	$V_3(l_1)$	600	–	–	–	–

Table 7.2 Index array for subgraphs G_1 and G_2

G_1	$V_0(l_0)$	$V_1(l_1)$	End
	0	2	3
G_2	$V2(l_0)$	$V_3(l_1)$	End
	0	1	1

The foreach statement in Line 15 is followed by a communication phase among subgraphs where the remote vertices communicate the values to the subgraph in which those vertices are local. In the example, communication of *dist* values of r_1 and r_2 in G_1 to l_0 and l_1 of G_2 (respectively) happens. The computation (*compu*) and communication (*comm*) in each iteration is shown in Table 7.3. Distributed computation requires the property values to be modified using commutative and associative operations. In this example, when the value of 500 communicated by vertex V_1 is received by vertex V_3 in phase 2, its own computed value of 675 is reduced by applying a MIN operation over these two values. This operation is typically sum, min, max, etc., and can also be a special function specified by the programmer.[1]

---

[1]More details can be found in the thesis: https://www.csa.iisc.ac.in/~srikant/papers-theses/Unnikrishnan-PhD-thesis.pdf.

Table 7.3 SSSP computation over two nodes

Iteration	$G_1(V_0, V_1)$ l_0, l_1, r_0, r_1	$G_2(V_2, V_3)$ l_0, l_1
$compu_0$	$0, \infty, \infty, \infty$	∞, ∞
$compu_1$	$0, 100, 75, \infty$	∞, ∞
$comm_1$	$0, 100, 75, \infty$	$75, \infty$
$compu_2$	$0, 100, 75, 500$	$75, 675$
$comm_2$	$0, 100, 75, 500$	$75, 500$

Table 7.4 Falcon data types

Data type	Description
Point	Vertex of a graph object
Edge	Edge of a graph object
Graph	The graph object. Consists of points and edges
Set	A static collection implemented as a *Union-Find* data structure
Collection	A dynamic collection with add and delete operations

7.3 Falcon Data Types

The data types provided by Falcon are listed in Table 7.4. Falcon provides Point, Edge, Graph, Set and Collection data types.

A Point data type stores the vertices of a graph object, and can have upto three dimensions with coordinate values stored in the fields x, y and z respectively.[2]

An Edge data type represents an edge and consists of *source*, and *destination* vertices, and also the *weight* of the edge, all of which can be accessed using *src*, *dst*, and *weight* fields of the Edge data type.

A Graph data type consists of vertices (Point) and edges (Edge). Additional properties can be associated with Graph, Point and Edge objects using the API functions of the Falcon data types. These functions are explained in Sect. 7.4.

A Set data type is implemented as a *union-find* data structure whose size (in terms of the number of primitive elements such as Point or Edge) cannot change at runtime. The two operations on a *Set* data type are the well-known operations on a *union-find* data structure, viz., *find* a primitive element in the *set* and perform a *union* of one subset with another disjoint subset. The parallel union-find data structure is explained in detail in Sect. 3.5.5.1. Falcon expects that *union()* and *find()* operations should not be called inside the same function because this may give rise to race conditions leading to incorrect results.

A Collection data type is implemented as a dynamic data structure. The *size* of a Collection object may be modified at runtime by *add* and *del* operations.

---

[2]Multi-dimensional representation is required for algorithms such as Delaunay mesh refinement.

The parallel implementation of the *add* and *del* operations uses atomic operations so that its sequential consistency is preserved.

7.4 Falcon API

The API functions of the Graph data type are shown in Table 7.5. The *read* function reads a graph object from an input file. The *addEdgeProperty* and *addPointProperty* functions add a new property to the edges and vertices respectively, of a graph object. The *addProperty* function adds a new property to a graph object. The *npoints* and *nedges* fields of the Graph data type store the number of vertices and edges respectively, of a graph object.

The important fields of the Edge data type are shown in Table 7.6. The *src* and *dst* fields store the source and destination vertices of an edge. The *weight* field stores the weight of an edge. The *del* function is used to delete an edge. Note that this function is different from the graph.delEdge() function. The former operates on an Edge and the latter operates on a Graph data type.

The important fields and functions of the Point data type are shown in Table 7.7. The Point data type can store int or float values and it can represent up to three dimensional coordinates. The values in each dimension can be accessed using the fields *x*, *y* and *z* respectively.

Table 7.5 Important API functions of the Falcon Graph data type

Function	Description
graph.read()	Read a Graph object
graph.addPointProperty()	Add a new property to each vertex
graph.addEdgeProperty()	Add a new property to each edge
graph.addProperty()	Add a new property to the Graph
graph.getWeight()	Get weight of an edge
graph.addPoint()	Add a new vertex
graph.addEdge()	Add a new edge
graph.delPoint()	Delete a vertex
graph.delEdge()	Delete an edge

Table 7.6 Important fields and functions of the Edge data type

Field	Type	Description
src	var	Source vertex of an Edge
dst	var	Destination vertex of an Edge
Weight	var	Weight of an Edge
del	Function	Delete an Edge

Table 7.7 Important fields and functions of the `Point` data type

Field	Type	Description
x, y, z	var	Stores `Point` coordinates in each dimension
getOutDegree()	Function	Returns number of outgoing edges of a `Point`
getInDegree()	Function	Returns number of incoming edges of a `Point`

7.5 Constructs for Parallel Execution and Synchronization

`Falcon` provides mutual exclusion via the `single` statement. The parallel constructs are the `foreach` statement and the `parallel sections` statement. The `Falcon` compiler supports reduction operations and does not support nested parallelism.

The `single` statement operates on either one item or a collection of items (see Table 7.8). The `else` block is optional. The first variant of the `single` statement is used to lock one item. In second variant, a thread tries to get a lock on a `Collection` object *coll* given as the argument. The lock on a collection of elements is required to implement dynamic algorithms where all the shared data among threads (e.g., a set of neighbor nodes) requires to be locked before participating threads can process the elements in the `Collection` object. A thread succeeds if it is able to lock all the elements in the `Collection` object. In both the variants, the thread that succeeds in acquiring a lock executes the code following it. The optional `else` block is executed by threads that fail to acquire the lock.

The `foreach` statement is an important parallel execution construct in `Falcon`. The `foreach` statement processes a set of elements in parallel. The `foreach` statement on objects of `Collection` or `Set` data type does not have an iterator. The *condition* and *advance_expression* are optional for both the variants of `foreach` statement (see Table 7.9). The `foreach` statement with the optional *condition* expression makes sure that only the elements in the *object*

Table 7.8 The `single` statement syntax and semantics

`single` (t1) { stmt block1 } `else` {stmt block2}	The thread that gets a lock on item t1 executes stmt block1, all other threads execute stmt block2
`single` (collection) { stmt block1} `else` {stmt block2}	The thread that gets a lock on all the elements in *collection* executes stmt block1, all others execute stmt block2

Table 7.9 The `foreach` statement syntax

`foreach` statement variant	Datatype
`foreach`(item (advance_expression) `In` object.iterator) (condition) { stmt_block }	`Point`, `Edge`, and `Graph`
`foreach`(item (advance_expression) `In` object) (condition) { stmt_block}	`Collection` and `Set`

Table 7.10 Iterators for the `foreach` statement

DataType	Iterator	Description
Graph	Points	Iterate over all points in graph
Graph	Edges	Iterate over all edges in graph
Graph	newppty	Iterate over all elements in the new property *newppty*
Point	nbrs	Iterate over all neighboring vertices
Point	innbrs	Iterate over *in_neighbours* of the point (for directed graphs)
Point	outnbrs	Iterate over *out_neighbours* of the point (for directed graphs)
Edge	nbrs	Iterate over neighboring edges
Edge	nbr1	Iterate over neighboring edges of Point P1 in Edge(P1, P2) (for directed graphs)
Edge	nbr2	Iterate over neighboring edges of Point P2 in Edge(P1, P2) (for directed graphs)

which satisfy the *condition* execute the *stmt_block*. The *advance_expression* is used to iterate from a given position instead of from offset zero (default). The $+(-)$ *advance_expression* makes the iterations go in the forward (reverse) direction, starting from the position specified in the *advance_expression*. The *object* used in the `foreach` statement can also be a dereferencing operator of a pointer to an object.[3] The `foreach` statement iterators and actions for different data types are briefly explained in Table 7.10.

Algorithm 7.2: `section` and `parallel sections` Syntax in `Falcon`

```
1 section {
2       statement_block
3 }
4 parallel sections {
5       one or more section statements
6 }
```

The `Graph` data type has the iterators *points* and *edges* to process all the vertices and edges respectively. It also has the fields *npoints* and *nedges* which contain the number of vertices and number of edges of the graph, respectively. The iterator `pptyname` is generated automatically by the `Falcon` compiler when a new property `pptyname` is added to a `Graph` object using the *addProperty()* function. For example, an iterator by the name *triangle* will be added to a `Graph` object when a property with name *triangle* is added to the graph object using the

---

[3]Boruvka's MST implementation in `Falcon` uses *advance_expression* and dereferencing of a pointer to an object in `foreach` statements.

addProperty() function. The number of elements of the property *triangle* is found using the field *ntriangle*. These fields are directly available to the programmer in Falcon programs after adding such a property. The iterators of *Point* and *Edge* data types are described in Table 7.10.

The outermost foreach statement is converted to appropriate parallel code depending on the target device. The parallel code has a global barrier after the parallel computation. This matches with the Bulk Synchronous Parallel (BSP) execution model. Nested inner foreach statements in Falcon are transformed to sequential for loops in the generated code.

7.5.1 section *and* parallel sections *Statements*

The syntax of the section and parallel sections statements is shown in Algorithm 7.2. All the section blocks inside the parallel sections statement run in parallel in the generated code. With this facility, Falcon can support multi-GPU systems and multiple CUDA kernels can run concurrently on different GPUs.

The code fragment in Algorithm 7.3 shows an example for parallel sections. There are two blocks of code enclosed in the two section blocks (Lines 22–28 and 15–21 respectively). The first and the second sections inside parallel sections call the functions *BFS* and *SSSP* (respectively) on the same input Graph object *graph*. The Falcon compiler for a multi-GPU machine generates code where each of these blocks of code gets executed on different GPUs concurrently. The total running time will be the maximum of the running times of SSSP and BFS. parallel sections reduces programming effort because there is no need to make two separate programs. It also improves throughput because overheads of loading the input graph twice and of other auxiliary processing are eliminated. The allocation of graph objects to different GPUs is taken care by the Falcon compiler [26].

7.6 Code Generation—Single Machine

Figure 7.2 shows the high level view of the code generation process for single machines adopted by the Falcon compiler. As an example, Algorithm 7.4 shows the optimized SSSP Falcon program. The program has three functions *reset()*, *update()*, and *relaxgraph()*. The Graph object *graph* is augmented with three Point properties *dist*, *olddist* and *uptd* of types int, int and bool, respectively (see Lines 18–20). Since we have explained SSSP in multiple contexts, only the points to be noted in the context of Falcon are mentioned here.

An important point to note in Algorithm 7.4 is that the code uses the graph only on the CPU. The programmer does not need to worry about two graph copies, or the

---

Algorithm 7.3: Multi-GPU BFS and SSSP in Falcon

```
1  SSSPBFS(char *name) { // begin SSSPBFS
2  Graph graph; //Graph object on CPU
3  graph.addPointProperty(dist, int);
5  graph.addProperty(changed, int); graph.addPointProperty(dist, int);
6  graph.getType() graph0, graph1;
7  graph.read(name); // read Graph from file to CPU
8  graph0 = graph; // copy entire Graph to GPU0
9  graph1 = graph; // copy entire Graph to GPU1
10 foreach(t In graph0.points) t.dist = 1234567890;
11 foreach(t In graph1.points) t.dist = 1234567890;
12 graph0.points[0].dist = 0;
13 graph1.points[0].dist = 0;
14 parallel sections { // do in parallel
15     section { // compute BFS on GPU1
16         while (1){
17             graph1.changed[0] = 0;
18             foreach(t In graph1.points) BFS(t, graph1);
19             if (graph1.changed[0] == 0) break;
20         }
21     }
22     section { // compute SSSP on GPU0
23         while (1){
24             graph0.changed[0] = 0;
25             foreach(t In graph0.points) SSSP(t, graph0);
26             if (graph0.changed[0] == 0) break;
27         }
28     }
29 }
31 } //end SSSPBFS
```

---

data transfer. All of these are taken care of automatically by the Falcon compiler, as explained in later sections.

Only the vertices which have the *uptd* property true will execute statements inside the *relaxgraph()* function as the foreach statement has a filter based on the condition (*t.uptd* == true) (see Line 27). The *update()* function checks whether *dist* value of each vertex has been reduced by *relaxgraph()* function using the condition $t.dist < t.olddist$ $\forall t \in graph.vertices$. The *uptd* property of a vertex is set to true if its *dist* value is reduced. Only such vertices participate in the next invocation of the *relaxgraph()* function. The foreach statement in Line 4 is not parallel as it is nested inside the foreach statement in Line 27 which calls the *relaxgraph()* function. The computation in the while-loop is the usual fixed-point computation. The library function MIN sets the variable *changed* to one, if reduction in *dist* value occurs, which results in the while-loop iterating once more. Otherwise, after the check at Line 28, the while-loop exits.

Algorithm 7.1 does not have a condition for the foreach statement which calls the *relaxgraph* kernel. This leads to $|E|$ number of atomic MIN operations in each

---

Algorithm 7.4: Optimized SSSP Code in `Falcon`

```
1   int changed = 1;
2   relaxgraph(Point p, Graph graph) {
3   |    p.uptd = false;
4   |    foreach( t In p.outnbrs ){
5   |    |    MIN(t.dist, p.dist + graph.getWeight(p, t), changed);
6   |    }
7   }
8   reset(Point t, Graph graph) {
9   |    t.dist = t.olddist = 1234567890;
10  |    t.uptd = false;
11  }
12  update(Point t, Graph graph) {
13  |    if( t.dist<t.olddist) t.uptd = true;
14  |    t.olddist = t.dist;
15  }
16  main(int argc, char *argv[]) {
17  |    Graph hgraph; // graph object
18  |    hgraph.addPointProperty(dist, int);
19  |    hgraph.addPointProperty(olddist, int);
20  |    hgraph.addPointProperty(uptd, bool);
21  |    hgraph.read(argv[1]);
22  |    foreach (t In hgraph.points) reset(t, hgraph);
23  |    hgraph.points[0].dist = 0; // source has dist 0
24  |    hgraph.points[0].uptd = true;
25  |    while( 1 ){
26  |    |    changed = 0; //keep relaxing on
27  |    |    foreach(t In hgraph.points) (t.uptd == true) relaxgraph(t, hgraph);
28  |    |    if (changed == 0) break; // reached fixed-point
29  |    |    foreach (t In hgraph.points) update(t, hgraph)
30  |    }
31  |    for (int i = 0; i <hgraph.npoints; ++i)
32  |         printf("i=%d dist=%d\n", i, hgraph.points[i].dist);
33  }
```

---

call to *relaxgraph* function. Thus Algorithm 7.1 reaches a fixed point with lesser number of iterations compared to Algorithm 7.4. Algorithm 7.4 iterates through those outgoing edges of the vertices $v \in G.V$ which have *uptd* property True. During the first iteration only the outgoing edges of the *source* vertex participate in the computation. This number increases as the wavefront of vertices advances. The advantage of Algorithm 7.4 is that the number of atomic MIN operations is smaller than that of Algorithm 7.1, leading to better performance, but with more iterations required to reach the fixed point.

The code generator of the `Falcon` compiler generates parallel C++ and CUDA codes for multi-core `CPU` and `GPU` respectively. The graph object is represented in Compressed Sparse Row (CSR) format by default, and Edge List format is also supported. The C++ classes *HGraph* and *GGraph* are used to store a graph object on multi-core `CPU` and `GPU` respectively, and both the classes inherit from the class *Graph*. The *Graph* class has a field named *extra* of type void∗ which is used to allocate extra properties added to a graph object using *addPointProperty*, *addEdgeProperty*, and *addProperty* methods. The Falcon compiler performs type checking and reports errors when there is mismatch in operators for domain-specific constructs. The Point and Edge variables are converted to integers.

7.6.1 Compilation of Data Types

The Point data is stored as a Union data type with float and int fields (*fpe* and *ipe* respectively). Point objects are stored in an array named *points* of size $|V|$. The fields of the Edge data type are stored in CSR format using two arrays *edges* and *index* of size $2 \times |E|$ and $(|V| + 1)$ respectively. Each edge stores two values, destination vertex and edge weight. The source vertex is computed from *index* array. A Graph object consists of collections of vertices and edges. The Set data type implementation uses the Union-Find data structure.[4] The Collection data type has been implemented using the *Thrust* library for GPU [4] and the *Galois Worklist*[5] for CPU [66].

7.6.2 Extra Property Memory Allocation

The number and types of extra properties vary across algorithms. The Falcon compiler generates code which allocates memory for the properties specified in the input program. Algorithm 7.5 shows the generated code for allocating the properties *dist*, *olddist* and *uptd* on multi-core CPU (Lines 6–11) and GPU (Lines 12–19). The properties are allocated in the *extra* field of the Graph class using the functions *malloc()* and *cudaMalloc()* on CPU and GPU respectively. The *extra* field of type void∗ allocated on GPU (device) cannot be accessed directly from the CPU (host). Therefore, it is allocated first and then copied to a temporary variable, *tmp*. Then the fields are allocated using the *tmp* variable.

7.6.3 Compilation of foreach Statement

A foreach statement is converted to a CUDA kernel for GPU and to an OpenMP pragma for CPU. Since foreach typically operates on a graph, it is useful to understand the implementation of a graph object stored in CSR format. A *graph* object has two arrays, *edges* and *index*. The *edges* array stores the destination vertex and edge weight of all edges in the graph, and is sorted by the source

---

[4]The details of implementation of Union-Find in Falcon are available in [26] and https://www. csa.iisc.ac.in/~srikant/papers-theses/Unnikrishnan-PhD-thesis.pdf.

[5]https://iss.oden.utexas.edu/?p=projects/galois.

Algorithm 7.5: SSSP: Generated Code for Extra Property Allocation on `CPU` and `GPU`

```
 1 struct struct_hgraph {
 2     int *dist, *olddist;
 3     bool *uptd;
 4 };
 5 struct struct_hgraph tmp;
 6 alloc_extra_graphcpu(HGraph &graph) {
 7     graph.extra = (struct struct_hgraph *) malloc(sizeof(struct struct_hgraph));
 8     ((struct struct_hgraph *)(graph.extra))→dist = (int *)
       malloc(sizeof(int)*graph.npoints);
 9     ((struct struct_hgraph *)(graph.extra))→olddist = (int *)
       malloc(sizeof(int)*graph.npoints);
10     ((struct struct_hgraph *)(graph.extra))→uptd = (bool *)
       malloc(sizeof(bool)*graph.npoints);
11 }
12 alloc_extra_graph(GGraph &graph) {
13     cudaMalloc((void **) &(graph.extra), sizeof(struct struct_hgraph));
14     cudaMemcpy(&tmp, (ep *)(graph.extra), sizeof(struct
       struct_hgraph),cudaMemcpyDeviceToHost);
15     cudaMalloc((void **) &(tmp.dist), sizeof(int)* graph.npoints);
16     cudaMalloc((void **) &(tmp.olddist), sizeof(int)* graph.npoints);
17     cudaMalloc((void **) &(tmp.uptd), sizeof(bool)* graph.npoints);
18     cudaMemcpy(graph.extra, &tmp, sizeof(struct struct_hgraph),
       cudaMemcpyHostToDevice);
19 }
```

---

vertex identifier. The number of outgoing edges from a vertex t is found using $(index[t + 1] - index[t])$. The starting offset for the outgoing edges from vertex t is $2 \times index[t]$. The GPU code for the *relaxgraph()* function in SSSP is shown in Algorithm 7.6 with the last two lines showing the CUDA kernel call to *relaxgraph()* from the host.

The code generated for (parallel) call to the *relaxgraph()* function for CPU is shown in Algorithm 7.7. The *relaxgraph()* function call inside the `foreach` statement is converted to an `OpenMP` parallel `for` loop by the `Falcon` compiler. The variable *TOT_CPU* carries a value equal to the number of cores in the CPU.

7.6.4 Data Transfer Between `CPU` and `GPU`

Data transfer between a `CPU` (host) and a `GPU` (device) is performed using the *cudaMemcpy()* function for pointer variables by specifying the direction as host-to-device (`cudaMemcpyHostToDevice`) or device-to-host (`cudaMemcpyDeviceToHost`). Algorithm 7.5 shows the code generated for allocation of pointer variables. The variable *changed* is assigned a value using *cudaMemcpyToSymbol()* and the value of the *changed* variable is read using

Algorithm 7.6: SSSP: GPU Version of the Code Generated for relaxgraph() and the Function Call from CPU

```
1  #define t ((( struct struct_hgraph *)(graph.extra)))
2  __global__ void relaxgraph( GGraph graph int x) {
3        int id = blockIdx.x * blockDim.x + threadIdx.x + x;
4        if( id <graph.npoints && t→uptd[id] == true ){
5              int falcft0 = graph.index[id];
6              int falcft1 = graph.index[id + 1] - graph.index[id];
7              for (int falcft2 = 0; falcft2 <falcft1; falcft2++) {
8                    ut0 = 2 * (falcft0 + falcft2); //edge index
9                    int ut1 = graph.edges[ut0].ipe; //dest point
10                   int ut2 = graph.edges[ut0 + 1].ipe; // edge weight
11                   GMIN( t->dist[ut1], t->dist[id] + ut2, changed);
12             }
13       }
14 }
15 int flcBlocks=(graph.npoints / TPB + 1)>MAXBLKS ? MAXBLKS : (graph.npoints / TPB
   + 1);
16 for (int kk = 0;kk<graph.npoints;kk += TPB * flcBlocks)
17       relaxgraph <<< flcBlocks, TPB >>>(graph, lev, kk);
```

Algorithm 7.7: CPU Version of the Code Generated for relaxgraph() and Its Call

```
1  #define t ((( struct struct_hgraph *)(graph.extra)))
2  void relaxgraph( int &p, HGraph &graph) {
3        if( id <graph.npoints &&t→dist[id] == true ){
4              int falcft0 = graph.index[id];
5              int falcft1 = graph.index[id + 1] - graph.index[id];
6              for (int falcft2 = 0; falcft2 <falcft1; falcft2++) {
7                    int ut0 = 2 * (falcft0 + falcft2); //edge index
8                    int ut1 = graph.edges[ut0].ipe; //dest point
9                    int ut2 = graph.edges[ut0 + 1].ipe; // edge weight
10                   HMIN( (t->dist[ut1], t->dist[id] + ut2, changed);
11             }
12       }
13 }
14 #pragma omp parallel for num_threads(TOT_CPU)
15 for (int i = 0; i <graph.npoints; i++) relaxgraph(i, graph);
```

cudaMemcpyFromSymbol() functions. Algorithms 7.8 and 7.9 show examples of code generated by the Falcon compiler for global variables.

7.6.5 Compiling single Statement

The single statement when used with only one element is converted to a compare-and-swap based code for CPU and GPU. When it is used with a collection of items,

---

Algorithm 7.8: Code Generated for Line 28:{`if (changed == 0) break;`} in Algorithm 7.4

---

1 int falctemp4;
2 cudaMemcpyFromSymbol((void *)&falctemp4, changed, sizeof(int), 0,
   cudaMemcpyDeviceToHost);
3 if (falctemp4 == 0) break;

Algorithm 7.9: Code Generated for Line 26: { `changed = 0;` } in Algorithm 7.4

---

1 int falcvt3 = 0;
2 cudaMemcpyToSymbol(changed, &(falcvt3), sizeof(int), 0, cudaMemcpyHostToDevice);

---

each thread tries to lock possible overlapping elements. If two or more threads try to lock collections of elements, say A and B, with common elements among them (i.e., $A \cap B \neq \Phi$), the thread with the lowest thread-id is given priority in locking its collection object. This is ensured by the `Falcon` code generator. The lock on a collection of elements requires a global barrier and the `Falcon` compiler implements a global barrier for GPU in software (see Algorithm 5.2 in Chap. 5).

7.6.6 Compiling Parallel Sections

The `parallel sections` block in `Falcon` is converted to an OpenMP *parallel sections* block. Each `section` in a `parallel sections` block will run as a separate thread. The `Falcon` compiler automatically assigns each section to a different device (CPU and multiple GPUs) so that all the computation happens in parallel.

7.6.7 Handling Reduction Operations

A reduction operation is converted to either a CUDA kernel or OpenMP code, based on the target device. The reduction operation *ReduxSum adds* all the elements which satisfy the (optional) condition. A single line of DSL code below carries out a reduction on the edges of the `Graph` object *graph*:

if (graph.edges.marked == `true`) mstcost ReduxSum= graph.edges.weight;

The *weight* of each edge in the *graph* whose *marked* property value is `true`, is added to the variable *mstcost*. This operation is used in the MST computation in `Falcon`. The code generated for the reduction operation on the `Graph` object which resides on GPU is shown in Algorithm 7.10. The CUDA kernel *RSUM0* which

---

Algorithm 7.10: Reduction Operation: Generated Code for GPU

```
1  #define DH cudaMemcpyDeviceToHost
2  #define HD cudaMemcpyHostToDevice
3  __device__ unsigned int dmstcost;
4  __global__ void RSUM0(GGraph graph, int FALCX) {
5      int id = blockIdx.x * blockDim.x + threadIdx.x + FALCX;
6      __shared__ volatile unsigned int reduxarr[1024];
7      if( id <graph.nedges ){
8          if(((struct struct_graph *)(graph.extra))→marked[id] == true)
9              reduxarr[threadIdx.x] = graph.edges[2 * id + 1];
10         else
11             reduxarr[threadIdx.x] = 0;
12         __syncthreads();
13         for (int i = 2; i <= TPB; i = i * 2) {
14             if (threadIdx.x == 0) reduxarr[threadIdx.x] += reduxarr[threadIdx.x + i / 2];
15             __syncthreads();
16         }
17         if (threadIdx.x == 0) atomicAdd(&dmstcost, reduxarr[0]);
18     }
19 }
20 // host (CPU) code
21 unsigned int hmstcost = 0;
22 cudaMemcpyToSymbol(dmstcost, &hmstcost, sizeof(unsigned int), 0, HD);
23 for (int kk = 0; kk <graph.nedges; kk += falcBlocks * TPB) {
24     RSUM0<<<falcBlocks, TPB>>>(graph, kk);
25     // falcBlocks is (graph.nedges ÷ TPB + 1).
26     cudaDeviceSynchronize();
27 }
28 cudaMemcpyFromSymbol(&hmstcost, dmstcost, sizeof(unsigned int), 0, DH);
29 mstcost = hmstcost;
```

---

performs the reduction, is called from the *host*. *RSUM0* uses a single-dimensional array *reduxarr* stored in the shared memory private to each streaming multiple processor (SM). The size of the array is 1024 and the CUDA kernel is called from the *host* (CPU) with the number of threads per block, $TPB \leq 1024$. The weights of the edges which have *marked* value `true` are added to *reduxarr* by each thread in the thread block (see Line 9). The values in the elements of the array *reduxarr* are added to $reduxarr[0]$ using a sequential `for` loop (see Lines 13–16). The local value computed by each thread block (available in $reduxarr[0]$ is then added to the global *device* (GPU) variable *dmstcost* (see Line 17). The final value of *dmstcost* is copied to the *host* variable *mstcost* (see Lines 21–29).

7.7 `Falcon` for Distributed Systems

The `Falcon` compiler supports distributed graph analytics on CPU clusters, GPU clusters using GPUs of each machine or both CPU and GPUs of each machine and multi-GPU machines. A programmer needs to write only a single program in `Falcon` and with appropriate command line arguments, the `Falcon` compiler generates programs for different distributed systems. The generated code uses Message Passing Interface (MPI) library calls for communication, and C++ (CUDA) code for CPU (GPU) device (respectively).

7.7.1 Storage of Subgraphs in `Falcon`

Real world graphs are very large in size and cannot be processed on a single machine. Typical examples are social network graphs such as Twitter and Facebook. Such graphs are partitioned into multiple subgraphs and each subgraph is processed on a separate machine. Distributed graph analytics in Falcon follows the BSP model (see Sect. 1.9.1.1) which involves a series of supersteps with each superstep having the three phases, parallel computation, communication, and synchronization. Graph analytics is efficient when there is proper balance in computation and communication. Computation is balanced if each subgraph requires a similar computation time during each superstep. Communication time is minimal if the number of edges across subgraphs is minimal. Optimum graph partitioning which balances partition sizes and minimizes the edge-cut across partitions is an NP-hard problem and needs to rely on heuristics (see Sect. 1.4).

A distributed system requires a different graph storage strategy than a shared memory system. The `Point` data type is associated with a *global vertex id* (GID) in the graph. The same vertex can have different *remote vertex id* (RID) in different subgraphs. Each vertex is also associated with a *local vertex id* (LID) in exactly one subgraph. Table 7.11 shows the partitioning of a graph with 15 vertices across three machines and remapping of GID to RID and LID. The GIDs of the vertices

Table 7.11 Conversion of GID to LID and RID on three machines M1, M2, M3

v_0	v_1	v_2	v_3	v_4	v_5	v_6	v_7	v_8	v_9	v_{10}	v_{11}	v_{12}	v_{13}	v_{14}
M1	$v_0(l_0)$		$v_1(l_1)$		$v_2(l_2)$		$v_3(l_3)$		$v_4(l_4)$	$v_6(r_5)$	$v_8(r_6)$	$v_9(r_7)$	$v_{11}(r_8)$	$v_{13}(r_9)$
M2	$v_5(l_0)$		$v_6(l_1)$		$v_7(l_2)$		$v_8(l_3)$		$v_9(l_4)$		$v_2(r_5)$	$v_4(r_6)$	$v_{11}(r_7)$	$v_{13}(r_8)$
M3		$v_{10}(l_0)$		$v_{11}(l_1)$		$v_{12}(l_2)$		$v_{13}(l_3)$		$v_{14}(l_4)$	$v_3(r_5)$	$v_5(r_6)$	$v_7(r_7)$	

Machine	0		1		2
Index	(M1)		(M2)		(M3)
Vertex offsets	(0, 5, 8, 10)		(0, 5, 7, 9)		(0, 5, 6, 8)

are $v_0, \ldots, v_{14}$. The first partition on machine M1 has ten vertices, of which v_0, v_1, v_2, v_3, v_4 are local to M1 (renamed as $l_0, \ldots, l_4$), v_6, v_8, v_9 are remote and located on machine M2 (renamed on M1 as r_5, r_6, r_7), and v_{11}, v_{13} are also remote and located on machine M3 (renamed as r_8, r_9). The offsets array (the last one in Table 7.11) shows the offsets of vertices for each machine. For example, the second block of offsets belongs to M2 and within that block, offsets 0, 5, and 7 indicate the starting offsets of local vertices on M2, remote vertices on M1, and remote vertices on M3, respectively. The last entry of each offset block (9 in the block for M2) indicates the number of vertices (local+remote) on a machine. It must also be noted that the number of remote vertices on a machine can be computed from the offset array. For example, offset$[1, 2]$—offset$[1, 1]$[6] gives the number of remote vertices on machine M2 which are local vertices on machine M1 (two in this example: v_2 and v_6).

In the edge-list format, each subgraph stores the outgoing edges with their LID as the source vertex. A vertex becomes remote when the destination vertex of an edge in the subgraph is not local to that subgraph. The role of the source vertex and the destination vertex are interchanged when the graph is stored in reverse edge-list format. Each edge is stored in exactly one subgraph as Falcon uses edge-cut partitioning. After every parallel computation phase, each subgraph gathers updated properties of the local vertices from subgraphs where its local vertex is a remote vertex. *Distributed computation requires vertex properties to be updated only using commutative and associative operations.* A programmer views a graph as a single object. Graph partitioning and distributed code generation are taken care of by the Falcon compiler.

Code generated by the Falcon compiler for distributed systems contains calls to MPI library functions such as *MPI_Isend()*,*MPI_Recv()*, *MPI_INIT*, *MPI_Comm_rank* etc. Important MPI library functions often present in the generated code are listed in Table 7.12. The generated code is compiled with a native machine compiler and the associated libraries (gcc, nvcc, etc.). Distributed execution with N processes assigns a unique integer value called *rank* to each process, where $0 \leq rank < N$. The *rank* value is used to uniquely identify each process. The communication between processes happens by specifying the *rank* of the processes.

Locking of elements in distributed systems is not supported by MPI. The Falcon compiler implements it in software using MPI library functions. The single construct uses distributed locking and it is transparent to the programmer.

7.7.2 Initialization Code

The number of processes with which the same binary is executed on multiple machines is specified as a command line argument during execution. Algorithm 7.11

---

[6]The offset array has been implemented as a single dimensional array in Falcon for efficiency reasons.

Table 7.12 MPI library functions used by the `Falcon` code generator

Function	Operation
MPI_INT	Initializes the MPI execution environment
MPI_Comm_rank	Used to find the rank of a process
MPI_Comm_size	Finds the number of processes involved in program execution
MPI_Isend	Send data from one process to another process
MPI_Recv	Receive data from a process
MPI_Get_count	returns the number of data elements received from the most recent `MPI_Recv()` operation
MPI_Barrier	Barrier for all the processes
MPI_Finalize	All the processes call this function before program termination

shows the code for initializing a distributed system with GPU. The total number of processes and the *rank* of the process are stored in the global variables *FALCsize* and *FALCrank* respectively. The value of *FALCrank* acts as a unique identifier for each process with different values in each process. It must be noted that each variable in a distributed system is duplicated with different memory locations in each process. Processes are very different from threads. The values are obtained by calls to the functions *MPI_Comm_size()* (see Line 6) and *MPI_Comm_rank()* (see Line 7) respectively. Communication is required for synchronizing subgraph properties. The variables *FALCrecvbuff* and *FALCsendbuff* (of type *FALCbuffer*) are used to receive and send data. Lines 12–19 allocate the buffer with target being a GPU device. The *dist* property in the SSSP computation is an example of a *property* to be communicated among processes running on different machines.

7.7.3 Allocation and Synchronization of Global Variables

Line 1 of Algorithm 7.4 declares a global variable *changed*. Algorithm 7.12 shows the generated code for different types of distributed systems (based on the value of the command line argument *TARGET*). For a target system with CPU and GPU devices, the variable *changed* will be duplicated to two copies, one each on CPU and GPU (Line 3, Algorithm 7.12).

Algorithm 7.13 shows how the `Falcon` code generator performs global variable allocation. It checks global variables which are read (use) or written (def) inside (possibly nested) `foreach` statements (parallel regions). Declaration statements appropriate to the target system (one for each device of the target) are generated for such global variables. Such global variables will be manipulated by several processes simultaneously and will require synchronization.

Algorithm 7.11: Initializing Distributed System: GPU Code

```
1   int FALCsize, FALCrank, temp;
2   MPI_Status *FALCstatus;
3   MPI_Request *FALCrequest;
4   void FALCmpiinit(int argc, char **argv) {
5       MPI_Init(&argc, &argv);
6       MPI_Comm_size(MPI_COMM_WORLD, &FALCsize);
7       MPI_Comm_rank(MPI_COMM_WORLD, &FALCrank);
8       FALCstatus = (MPI_Status *)malloc(sizeof(MPI_Status) * FALCsize);
9       FALCrequest = (MPI_Request *)malloc(sizeof(MPI_Request) * FALCsize);
10      gethostname(FALChostname, 255);
11      cudaMalloc(&FALCrecvbuff, sizeof(struct FALCbuffer ));
12      cudaMalloc(&FALCsendbuff, sizeof(struct FALCbuffer )*FALCsize);
13      cudaMalloc(&FALCrecvsize, sizeof(int) * FALCsize);
14      cudaMalloc(&FALCsendsize, sizeof(int) * FALCsize);
15      for (int i = 0; i <FALCsize; i++) {
16          temp = 0;
17          cudaMemcpy(&FALCrecvsize[i], &temp, sizeof(int),
                cudaMemcpyHostToDevice);
18          cudaMemcpy(&FALCsendsize[i], &temp, sizeof(int),
                cudaMemcpyHostToDevice);
19      }
20  }
21  int main( int argc, char *argv[]) {
22      ......//code not relevant for distributed execution.
23      FALCmpiinit(argc, argv);
24      ....
25  }
```

Algorithm 7.12: Generated Code for Line 1 (int changed;) in Algorithm 7.4 for Different Targets

---

```
1   int changed; // TARGET: CPU, and CPU cluster
2
3   __device__ int changed; // TARGET: GPU, Multi-GPU, and GPU cluster
4
5   __device__ int changed; int FCPUchanged; // TARGET: CPU+GPU cluster
```

7.7.4 Distributed Communication and Synchronization

The Falcon compiler achieves synchronization of a global variable *var* in a distributed system generating code using the code pattern in Algorithm 7.14.

An example how the global variable *changed* is synchronized is shown in Algorithm 7.15. Synchronization is required after a modification of the variable in parallel computations on several nodes. The local value of the variable *changed* is sent by each process, to all other processes, using an asynchronous $MPI\_ISend()$ function (Lines 6–8). The value of the *changed* variable received from all other

---

Algorithm 7.13: Global Variable Allocation by the Falcon Compiler for a Distributed System

---

```
1 for (each parallel_region p in program) {
2    for (each var in globalvars) {
3       if (def(var, p) or use(var, p))
5          allocate var on target device / devices
6    }
7 }
```

Algorithm 7.14: Pseudo Code for Synchronizing a Global Variable *var*

```
1 for (each remote node i) sendtoremotenode(i, var);
2 Type tempvar = 0; // Type is Data type of var
3 for (each remote node i) {
4    receive from remote node(i, tempvar);
5    update var using tempvar
6    based on function used to modify var. (MIN, MAX, ADD, etc.).
7 }
```

Algorithm 7.15: CPU Cluster: Code for Synchronizing the Global Variable *Changed*

```
1 #define MCW MPI_COMM_WORLD
2 #define MSI MPI_STATUS_IGNORE
3 MPI_Status status[8]; // 0 ≤ rank, NPARTS < 8
4 MPI_Request request[8];
5 int falctv4;
6 for (int i = 0; i <NPARTS; i++) {
7    if (i != rank) MPI_Isend(&changed, 1, MPI_INT, i, messageno, MCW, &request[i]);
8 }
9 for (int i = 0; i <NPARTS; i++) {
10   if (i != rank) MPI_Recv(&falctv4, 1, MPI_INT, i, messageno, MCW, MSI);
11   changed = changed + falcvt4;
12 }
13 if (changed == 0)  break;
```

---

processes is collected in the temporary variable *falctv4*. The value is collected using the synchronous $MPI\_Recv()$ function. The received value is used to update the variable *changed* (Line 11). If the value of the variable *changed* is zero, the break statement is executed (Line 13).[7]

The code in Algorithm 7.15 is for a CPU cluster with eight machines. The code for a GPU cluster will have additional code for copying variables from and to the GPU. The Falcon compiler uses OpenMPI library with *cuda-aware-mpi* support for multi-GPU machines.

---

[7]This statement is generated if it is a part of the source code as in the SSSP or other algorithms.

7.7.5 Code Generation for Set and Collection Data Types

Algorithm 7.16: Distributed Union in Falcon

```
 1  if( rank(process) != 0 ){
 3  |    add each union request Union(u, v) to the buffer
 5  |    Send the buffer to the process with rank == 0
 7  |    receive parent value from node zero
 9  |    update local set
10  }
11  if( rank(process) == 0 ){
13  |    receive Union(u, v) request from remote node (process)
15  |    perform union; update parent of each element
17  |    send parent value to each remote node (processes)
18  }
```

Falcon implements distributed Union-Find Set data type on top of the Set implementation for single machines [26]. The distributed Set uses the first process ($rank = 0$) for collecting *union()* requests from all the other processes. The processs with *rank* zero performs the *union()* operation and sends the updated parent value to all the other processes involved (see Algorithm 7.16). This seems to be adequate for small clusters that are tightly coupled. A more complex union would be warranted for large clusters.

Algorithm 7.17: Collection Synchronization in Falcon

```
 1  foreach( item in Collection ){
 3  |    if (item.master-node≠rank(process_in_node))
 5  |        add item to buffer[item.master-node] and delete item from Collection
 6  }
 8  foreach ( i ∈ remote-node) send buffer to remote-node(i)
10  foreach ( i ∈ remote-node) receive buffer from remote-node(i)
11  foreach( i ∈ remote-node ){
12  |    foreach( j ∈ buffer[i] ){
14  |    |    update property values using buffer[i].elem[j]
16  |    |    addtocollection(buffer[i].elem[j])
17  |    }
18  }
```

The Collection data type supports duplicate elements. The *add()* function of the Collection data type is overloaded and supports adding elements to a Collection object with no duplicate elements. This avoids sending the same data of remote nodes to the corresponding master nodes multiple times. A

Algorithm 7.18: Falcon Program with single Statement

```
 1  fun(Point p, Graph g {
 2  |    Program segment 1. //Does not contain any single statement.
 3  |    foreach( t In p.outnbrs ){
 4  |    |    if( single(t.lock) ){
 5  |    |    |    {Stmt_Block}
 6  |    |    }
 7  |    }
 8  |    Program segment 2. //Does not contain any single statement.
 9  }
10  main() {
11  |    ...
12  |    //graph is the graph structure being processed.
13  |    foreach( p In graph ){
14  |    |    fun(p, graph);
15  |    }
16  |    ...
17  }
```

---

global Collection object is synchronized by sending remote elements in the Collection object to the appropriate master mode. The Collection object is synchronized as shown in Algorithm 7.17.

7.7.6   Code Generation for Parallel and Synchronization Statements

We now explain the code generation for single and foreach statements.

7.7.6.1   single Statement

The generated (distributed) code of a Falcon program operates on each subgraph G_i of G using a separate process. Only one thread among all the threads on all the nodes should succeed in getting a *lock* on an element or a collection of elements. The code generated by the Falcon compiler ensures that a function with a single statement is executed in two phases and that the semantics is preserved. An example Falcon source code with single statement is shown in Algorithm 7.18. The pseudo-code for the generated code (common for both CPU and GPU) is shown in Algorithm 7.19.

In the first phase, each node acquires a local lock on its vertices (local and remote). The generated code then performs communication with the process of

---

Algorithm 7.19: Generated Pseudo-Code for `single` Statement (for Algorithm 7.18)

```
 1  //Generated code (pseudo-code) for node i processing subgraph G_i.
 2  fun1(Point p, Graph g) {
 3      Program segment 1. //Does not contain any single statement.
 4      foreach( t In p.outnbrs ){
 5          CAS(t.lock, MAX_INT, rank)
 6      }
 7  }
 8  fun2(Point p, Graph g {
 9      foreach( t In p.outnbrs ){
10          if( CAS(t.lock, rank, MAX_INT - 1) ){
11              {Stmt_Block}
12          }
13      }
14      Program segment 2. // Does not contain any single statement.
15  }
16  main() {
17      ...
18      // Reset locks.
19      foreach(Point t In G_i.points) t.lock = MAX_INT;
20      // Global lock acquisition protocol.
21      //Step (a) Acquire local locks.
22          foreach(Point p in G_i.points) fun1(p, G_i);
23          // Send local lock values to Process 0.
24          foreach(Point t In G_i.points){
25          if (t.lock ≠ MAX_INT) {
26              send t.lock to process with rank == 0.
27          }
28          }
29      //Step (b) Send MIN values
30      if (process.rank == 0){
31          // Note that all points in G are processed, not only G_i.
32              foreach(t In G.points){
33              t.lock = MIN  of all lock values received for t.
34              Send t.lock value to all remote nodes of t.
35              }
36      }
37      (c) update lock value locally
38          foreach(Point p In G_i.points){
39          if (process.rank > 0) {
40              Receive p.lock value from process zero.
41              Update own p.lock value.
42          }
43          }
44      // Now execute code under lock.
45      foreach(Point p In G_i.points) fun2(p, G_i);
46      ...
47  }
```

rank zero. Processes with *rank* greater than zero send all the lock values which are modified from the initial value of *MAX_INT*, to the process with *rank* zero (P_0). The process P_0 collects the *lock* values from the remote nodes and sets the *lock* value for all the points to the smallest *rank* value. The process P_0 then sends the modified *lock* value back to each remote node. In the second phase, the remote nodes receive the final *lock* value for each vertex from process P_0, and update the *lock* value of each of their vertices. The node which obtains the lock for vertex i executes the body of the `single` statement for vertex i.

The pseudo code for distributed *single* statement is shown in Algorithm 7.19. Function *fun()* is duplicated into two versions, *fun1()* and *fun2()*. The function *fun1()* has {Program segment 1} and ends with a CAS operation replacing the *single* statement that acquires local locks. Function *fun2()* begins with a CAS operation replacing the *single* statement that operates on the acquired lock (if any). Function *fun2()* also has *stmt_block*{} of the `single` statement and ends with Program segment 2. The calling portion of function *fun()* contains the code for acquiring the global lock for each vertex. Such an implementation is used in the Boruvka's-MST implementation, wherein different threads race for operating on a component.

Algorithm 7.20: CPU Cluster: Prologue and Epilogue Code for *relaxgraph* Call

```
 1  // prefix-code
 2  #pragma omp parallel for num_threads(FALC_THREADS)
 3  for (int i = graph.nlocalpoints; i <graph.nremotepoints; i++) {
 4  |   tempdist[i] = ((struct struct_graph *)(graph.extra)) ->dist[i];
 5  }
 6  #pragma omp parallel for num_threads(FALC_THREADS)
 7  for (int i = 0; i <graph.nlocaledges; i++) {
 8  |   relaxgraph(i, graph);
 9  }
10  // suffix-code
11  for (int kk = 1; kk <FALCsize; kk++) {
12  |   #pragma omp parallel for num_threads(FALC_THREADS)
13  |   for (int i = graph.offset[kk - 1]; i <graph.offset[kk]; i++) {
14  |   |   addto_sendbuff(i, graph, FALCsendsize, FALCsendbuff, kk - 1);
15  |   }
16  }
```

7.7.7   *foreach Statement*

The *relaxgraph()* function of the SSSP program (see Algorithms 7.1 and 7.4) updates the *dist* value of the destination vertex (t) of an edge $e : p \rightarrow t$ using a MIN function and the vertex t could be a remote vertex for a subgraph $G_i \subset G$. The Falcon compiler generates prologue and epilogue codes for a parallel `foreach` statement, if the statement body updates a remote vertex. Such a code fragment using OpenMP is shown in Algorithm 7.20.

Algorithm 7.21: SSSP CPU Cluster: Update *dist* Property

```
 1 #define MCW MPI_COMM_WORLD
 2 #define MSI MPI_STATUS_IGNORE
 3 #define graphep ((struct struct_graph * )(graph.extra)) int totsend = 0;
 4 for (int i = 0; i <FALCsize; i++) {
 5     if( (i != FALCrank) ){
 6         totsend += sendsize[i];
 7         MPI_Isend((sendbuff[i].vid), sendsize[i], MPI_INT, i, messageno, MCW,
                &request[i]);
 8         MPI_Isend((sendbuff[i].dist), sendsize[i], MPI_INT, i, messageno + 1, MCW,
                &request[i]);
 9     }
10 }
11 for (int kk = 0; kk <(FALCsize); kk++) {
12     if( (kk != FALCrank) ){
13         MPI_Recv(recvbuff[0].vid, graph.hostparts[0].npoints, MPI_INT, i, messageno,
                MCW, &FALCstatus[i]);
14         MPI_Recv(recvbuff[0].dist, graph.hostparts[0].npoints, MPI_INT, i, messageno +
                1, MCW,MSI);
15         MPI_Get_count(&FALCstatus[kk], MPI_INT, &FALCnamount);
16         #pragma omp parallel for num_threads(FALC_THREADS)
17         for (int i = 0; i <FALCnamount; i++) {
18             int vertex = FALCrecvbuff[0].vid[i];
19             if(graphep - >dist[vertex] >FALCrecvbuff[0].dist[i])
20                 graphep - >dist[vertex] = FALCrecvbuff[0].dist[i];
21         }
22     }
23 }
```

The code first copies the current *dist* value of remote vertices to a temporary buffer *tempdist* (Line 4). Remote vertices have indices in the range: [G_i.localpoints, G_i.remotepoints] for any $G_i \subset G$. The *relaxgraph()* function is called after the copy operation (Line 8). The number of *remote vertices* for a remote node (machine) with index kk is ($offset[kk] - offset[kk - 1]$). The *add_to_sendbuff()* function (Line 14) checks for the condition ($tempdist[i]! = dist(i)$) and adds the *remote vertices* of the subgraph G_{kk} which satisfy this condition, to $FALCsendbuff[kk]$. $FALCsendbuff[kk]$ contains a set of tuples ($dist, loca\_id$) for the subgraph G_{kk}, where $local\_id$ is the local vertex id (LID) of a vertex in the subgraph, where the remote vertex is a local vertex. Then these buffer values are sent to the respective remote nodes. Node p receives ($dist, local\_id[p]$) pairs from all the remote nodes and updates its own *dist* value, by taking the MIN of its own *dist* value and the remote values. Algorithm 7.21 shows details of how this is achieved using MPI library calls.

Chapter 8
Experiments, Evaluation and Future Directions

This chapter discusses a summary of experimental results of running elementary graph algorithm implementations on multi-core CPU, GPU, and distributed systems for different types of graph inputs. A description of the future directions for research in distributed graph analytics is also provided for interested readers who want to probe further.

8.1 Introduction

The performance of different types of algorithms varies depending on the characteristics of both the machines and the graph inputs, apart from the strategy of implementation. The results of running several benchmarks on different types of machines and also different types of graph inputs are now presented. While there is no set of rules that can guide a programmer on the choice of implementation and machine, it is worthwhile to pay attention to warp divergence, graph partitioning and communication overhead before choosing the platform and the implementation strategy. The running times of different Falcon DSL implementations for the same benchmarks vary. Detailed results for different benchmarks are available in [26, 27].

8.2 Single Machine Experiments

A machine with a Xeon(R)W-2123 processor with eight cores running at a frequency of 3.6 GHZ, with 8 MB private cache per core, a volatile memory of 32 GB, and an Nvidia GeForce RTX-2080 GPU, was used for the experiments.

© Springer Nature Switzerland AG 2020
U. Cheramangalath et al., *Distributed Graph Analytics*,
https://doi.org/10.1007/978-3-030-41886-1_8

Table 8.1 Road networks

Input	$|V|$	$|E|$
USA-E	3,598,623	8,778,116
USA-W	6,262,104	15,248,146
USA-CTR	14,081,816	34,292,496
USA-USA	23,947,347	58,333,344

Table 8.2 Random graphs

Input	$|V|$	$|E|$
rand8M	8M	24M
rand16M	16M	64M
rand32M	24M	96M
rand48M	32M	128M
rand64M	64M	256M

Table 8.3 R-MAT graphs

Input	$|V|$	$|E|$
rma10M	10M	24M
rmat20M	20M	64M
rmat40M	40M	96M
rmat60M	60M	128M
rmat80M	80M	256M

The GPU has 4352 streaming processors (SPs) with 68 streaming multiprocessors (SMs) and 64 SPs per SM. It also has 11 GB of volatile memory and SPs run at a frequency of 1350 MHz.

The inputs used for evaluation consist of random graphs, RMAT graphs, and road graphs. The inputs are listed in Tables 8.1, 8.2, and 8.3 (M denotes Million). The random graphs were created using the graph generator tool available in the Galois [66] framework. The RMAT graphs were created using the *GTGraph* tool with default values for parameters a, b, c, d [154] which decide the graph sparsity and edge distribution. The road network graphs are USA road networks available in public domain [155]. The benchmarks include Breadth-First Search (BFS), Single Source Shortest Paths (SSSP), Connected Components using Union-Find (CC), and Boruvka's Minimum Spanning Tree (MST) algorithms programmed in Falcon.

The baseline running times for SSSP and BFS are listed in Table 8.4 for GPU and in Table 8.5 for CPU. The baseline BFS and SSSP algorithms use the atomic MIN library function. Both the algorithms use only one vertex (*dist*) property to store the BFS and the SSSP distances. The maximum BFS distance for R-MAT graph is ∞ as the graph is not fully connected. The number inside parentheses (after ∞) for the R-MAT graphs is the number of iterations taken to reach the fixed point. The main kernel which computes the BFS and the SSSP distances, runs for all the vertices in the graph object. The optimized version of Falcon program for BFS uses a level-based traversal with no atomic operations. The one for SSSP uses three vertex properties *dist*, *olddist* and *updated*, and one additional kernel.

Table 8.4 GPU running time

Input	Time (ms)		BFS distance
	BFS	SSSP	
rand8M	94	178	18
rand16M	238	416	20
rand24M	367	615	19
rand32M	467	870	18
rand48M	764	1340	19
rand64M	950	1595	26
rmat10M	149	571	∞ (5)
rmat20M	335	1490	∞ (5)
rmat30M	530	2655	∞ (4)
rmat40M	185	305	∞ (9)
rmat60M	252	444	∞ (16)
rmat80M	1540	7033	∞(8)
USA-E	505	900	2878
USA-W	903	1451	3137
USA-CTR	5972	11,938	3826
USA-USA	6440	12,124	6261

Table 8.5 CPU running time

Input	Time (ms)		BFS distance
	BFS	SSSP	
rand8M	1680	2658	18
rand16M	3440	5968	20
rand24M	5743	9235	19
rand32M	7755	13,890	18
rand48M	12,502	19,327	19
rand64M	13,993	20,893	26
rmat10M	2251	13,602	∞ (5)
rmat20M	5180	17,107	∞ (5)
rmat30M	6282	31,811	∞ (4)
rmat40M	3214	4842	∞ (9)
rmat60M	4862	7470	∞ (16)
rmat80M	22,249	83,937	∞(8)
USA-E	11,359	29,826	2878
USA-W	23,727	58,135	3137
USA-CTR	180,603	377,313	3826
USA-USA	185,661	646,217	6261

The number of iterations of the BFS kernel for R-MAT graphs was below ten. The average number of vertices that participated in an iteration is very high due to the low diameter of these graphs. This is good enough for utilization of the 4352 GPU cores.

The four road network graphs have high diameters (BFS distance) which range from 2878 to 6261. This leads to less parallelism per iteration as the number of iterations for the BFS and the SSSP computation is directly related to the BFS distance. The running times for road network graphs are high on GPU and CPU as given in Tables 8.4 and 8.5 respectively. Road network graphs have a poor performance on GPU and Δ-stepping computation is more suitable for road network graphs on CPU.

CUDA runtime performs *warp-voting* for the SIMT Nvidia-GPU, which helps in reducing the running time of level-based BFS. The level based BFS computation has a conditional block with the condition ($t.dist == level$) in the kernel. If none of the threads in a warp satisfies the condition, all the thirty two threads finish execution. The naïve BFS algorithm has no such condition and uses an atomic MIN operation, which leads to its high running time. The SSSP computation also benefits by *warp-voting* with the condition ($t.updated == True$). This condition makes sure that only the vertices whose distance was reduced in the previous iteration take part in the current iteration.

8.2.1 Random and R-MAT Graphs

The speedup of optimized algorithms over naïve algorithms on random graphs and RMAT graphs are shown for GPU and CPU in Figs. 8.1 and 8.2 respectively. The speedup depends on graph topology, presence of atomic operations, the way threads are scheduled at runtime, etc. We observe that GPU programs are much faster than the CPU versions. The multi-core CPU used for evaluation has only 8 computing cores while the number of SPs for the GPU is 4352. The performance is likely to be better on multi-core CPUs with more number of cores.

OpenMP based code is not always good for a multi-core CPU. However, the Δ-stepping algorithm gets benefited considerably. Figure 8.3 shows the speedup of the Δ-Stepping algorithm over naïve Falcon code for random and RMAT graphs with BFS and SSSP benchmarks.

8.2.2 Road Network Graphs

The road networks exhibit poor performance when implemented using OpenMP for BFS and SSSP computations. In contrast, the Δ-stepping algorithm runs extremely fast for road networks [26] on CPU. Figure 8.4a shows the speedup of optimized Falcon OpenMP code over naïve Falcon code. The Δ-stepping SSSP and BFS are much faster than the respective naïve programs as shown in Fig. 8.4b. Figure 8.5 shows the speedup of optimized Falcon BFS and SSSP over naïve BFS and SSSP respectively.

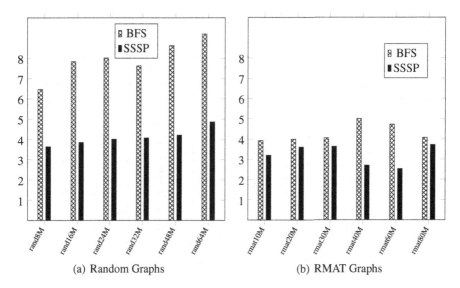

Fig. 8.1 GPU: speedup of BFS and SSSP (optimized `Falcon` code) over Naïve `Falcon` code

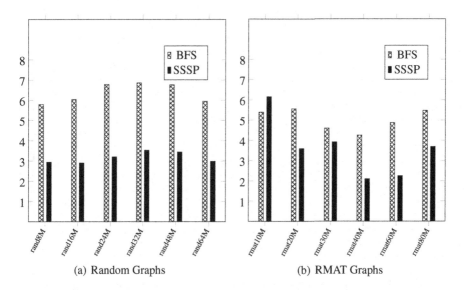

Fig. 8.2 CPU: speedup of BFS and SSSP (optimized `Falcon` code) over naïve `Falcon` code

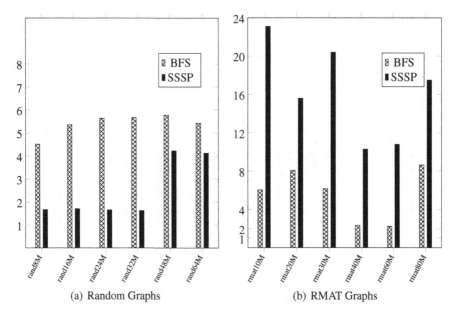

Fig. 8.3 CPU: speedup of Δ-stepping BFS and SSSP `Falcon` code over Naïve Falcon code

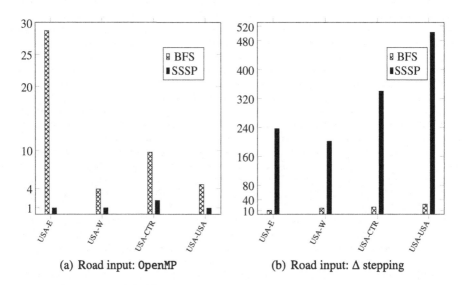

Fig. 8.4 CPU: speedup of BFS and SSSP (optimized `Falcon` code) over Naïve `Falcon` code

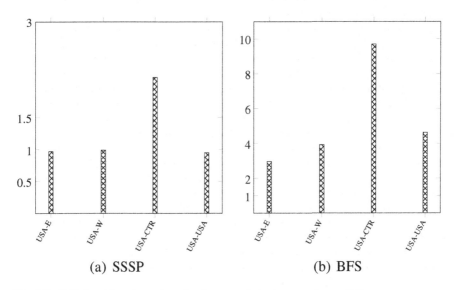

(a) SSSP   (b) BFS

Fig. 8.5 GPU: (road input) speedup of `Falcon`-optimized over `Falcon`-Naïve

8.2.3 Connected Components and MST

The running times of the CC and the MST algorithms for CPU (8-core) are shown in Table 8.6 and for GPU are shown in Table 8.7. The GPU exploits parallelism well. The road network graphs are not affected by the diameter in these benchmarks, as path compression in the Union-Find algorithm happens quickly. These algorithms are not traversal algorithms like SSSP and BFS.

Table 8.6 MST and CC on CPU	Benchmark	Input	Time (ms)
	MST	rand48M	19,151
		rand64M	21,100
		rmat60M	7936
		rmat80M	8746
		USA-CTR	782
		USA-USA	999
	CC	rand48M	2618
		rand64M	2949
		rmat60M	2708
		rmat80M	2618
		USA-CTR	88,038
		USA-USA	110,526

Table 8.7 MST and CC on GPU

Benchmark	Input	Time (ms)
MST	rand48M	2118
	rand64M	2139
	rmat60M	764
	rmat80M	776
	USA-CTR	119
	USA-USA	176
CC	rand48M	260.5
	rand64M	383
	rmat60M	162
	rmat80M	120
	USA-CTR	3129
	USA-USA	4281

8.2.4   Summary

The performance of an algorithm depends on many factors including graph diameter, degree distribution, warp divergence (in GPU), presence of atomic operations etc., in additional to the program. The storage of graphs for coalesced access is also important. Road networks have very low parallelism and are not suitable for GPUs. The Δ-stepping algorithm is suitable for processing road networks on multi-core CPUs. Efficient implementation of data structures is also crucial for performance. For example, parallel Union-Find needs to be efficient for Boruvka's MST computation.

8.3   Distributed Systems

Distributed execution has overheads of communication and storage space for remote vertices of subgraphs introduced by graph partitioning. The communication volume can be reduced by sending only the required data. Falcon uses vertex-cut partitioning. This section emphasizes results on communication overhead. The graphs used are different from the ones used for single machines. They are much larger and therefore, cannot be fitted inside the memory of a single machine.

For distributed execution experiments, we used a multi-GPU machine, a GPU cluster and a CPU cluster:

- The multi-GPU machine used for evaluation had eight GPUs, each being an Nvidia-Tesla K40 GPU with 2880 cores and 12 GB volatile memory. The machine had Intel(R) Xeon(R) multi-core CPU as the host, with 32 cores and 100 GB volatile memory.

Table 8.8 Inputs used for distributed processing

Input	Type	|V|	|E|
rmat100M	Random	100M	1000M
rmat200M	rmat	200M	200M
uk-2005 [156]	Web	39,460,000	921,345,078
uk-2007 [156]	Web	105,896,555	3,738,733,602
Twitter [157]	Social n/w	41,652,230	1,468,364,884

- The CPU cluster used for the experiments had sixteen machines. Each machine of the cluster consisted of two CPU sockets with 12 Intel Haswell cores operating at 2.5 GHz and 128 GB volatile memory.
- The GPU cluster had eight nodes, each with an Nvidia-Tesla K40 GPU card with 2880 cores and 12 GB volatile memory. Each node had a 12 core multi-core CPU as its host.

The benchmarks were run in a distributed fashion with the graph inputs partitioned across the machines or devices. Graph inputs of bigger size were used. The inputs used for evaluation are shown in Table 8.8. They consist of RMAT, web, and social network graphs. Falcon programs follow the BSP execution model.

8.3.1 Multi-GPU

Eight GPUs were organized into two clusters with four GPUs in each cluster. The GPUs in each cluster can communicate directly (without involving the host). This feature is enabled in CUDA using the *peer-access* API functions.

Table 8.9 shows the running times of different benchmarks (SSSP, BFS, CC) on the multi-GPU machine with increasing order of number of GPUs. The values of

Table 8.9 Multi-GPU: running time of SSSP, BFS and CC on public inputs

Input	Num. of GPUs	Time (ms)		
		SSSP	BFS	CC
UK-2005	2	8408	611	1267
	4	5457	572	1009
	6	5524	875	945
	8	5273	744	939
UK-2007	4	13013	1540	7201
	6	11864	1397	6910
	8	10780	1311	6791
Twitter	2	6354	1001	8095
	4	3950	581	4410
	6	3089	491	3507
	8	2674	945	2689

$|E|/|V|$ are 23, 35, and 35 for UK-2005, UK-2007 and Twitter inputs respectively. This is much higher than the value of ten used in RMAT inputs. The values shown in the table are best among edge based and vertex based computations. The Twitter input has a high variance in degree and it was observed that vertex based computation is upto 7 times slower than edge based computation. Variance in degree distribution results in thread divergence in GPUs. The Falcon compiler used OpenMPI library with *cuda-aware-mpi* to reduce communication overhead between GPUs on the same machine.

8.3.2 GPU Cluster

Computation on a GPU cluster has a high communication overhead. The communication from a GPU on machine M_i to a GPU on machine M_j is required to go through several stages. The data to be communicated is in the volatile memory of GPU in M_i. The data is first copied to the volatile memory of the host CPU in M_i, and then it is sent over the network to the volatile memory of the host CPU in machine M_j. Then it gets copied to the GPU memory of machine M_j. The BFS algorithm spends more than 90% of the time in communication on a GPU cluster with 8 nodes.

8.3.3 CPU Cluster

The computation time is high for CPUs compared to massively parallel GPUs, and the total time spent for communication on CPU clusters is very less when compared to the computation time [27].

8.3.4 CPU + CPU Cluster

The CPU+GPU cluster uses both the CPU and the GPU of each machine for computation. The BSP model imposes a barrier after computation and communication. The computation time is high for CPUs compared to massively parallel GPUs. The communication time is low in CPU-CPU communication compared to GPU-GPU communication. This imbalance in communication and computation results in making the CPU+GPU cluster slower than all other distributed machine combinations.

Table 8.10 Running time (in seconds) of RMAT graphs on eight GPUs or four CPUs and GPUs

Input	Algorithm	Multi GPU	CPU cluster	GPU + CPU cluster
rmat100M	BFS	1.2	3.7	8.4
rmat100M	SSSP	5.0	9.3	31.6
rmat100M	CC	1.1	2.7	81.3
rmat200M	BFS	1.8	6.9	16.7
rmat200M	SSSP	10.9	18.1	66.9
rmat200M	CC	2.4	5.7	17.4

8.3.5 Results

Table 8.10 compares the running times of multi-GPU, GPU cluster and CPU+GPU cluster. The multi-GPU machine with *peer-access* capability is always faster. The running time on the GPU cluster increases considerably due to the communication overhead. The GPU+CPU cluster takes longer due to imbalance in computation time and communication time in CPU and GPU. Detailed results can be found in [27].

8.4 Future Directions in Distributed Graph Analytics

Graph analytics has witnessed a rapid growth in the recent past. This is a clear indicator of its importance, and is due to more and more applications being modeled using graphs. Entities and their relationships can be naturally modeled as graph structures. Well-known algorithms can then be readily applied to these problems. Scalability has been, currently is, and will definitely continue to be at the forefront of graph analytics research. Several aspects of graph processing are either still unexplored or are in their infancy. We envision that the following areas would witness acceleration in the days to come.

8.4.1 Custom Graph Processing

Architectural advances have been fueling cutting-edge research in compilers, programming languages, as well as systems. Datacenters and other large hardware installations house a variety of devices, including multiple types of shared-memory multi-processors, asymmetric multi-core processors having the same ISA, GPUs with varying compute capabilities, TPUs, FPGAs, ASICs, etc. All these different devices are connected to each other via a multitude of network topologies. It is imperative, as well as non-trivial to generate efficient customized code for an individual device, a subset of the devices, or for all the network devices for the same graph application. Depending upon the device characteristics, their performance toward graph processing, and hardware availability, users may want to mix-and-match devices against a performance criteria—which could be execution time, energy efficiency, performance per watt per dollar, etc. Domain-agnostic and domain-specific languages need to evolve API and constructs to help users achieve such a complex goal. Compilers also need to be able to generate a variety of backend codes from the same algorithmic specification [158].

8.4.2   Adaptive Graph Partitioning and Processing

While graphs can be categorized based on their structural characteristics, very large graphs exhibit heterogeneity also within themselves. For instance, some vertices have low clustering coefficient, while some have it very high, and many hover around average. Most of the graph processing systems treat different parts of the graph in a uniform manner. However, we need different types of processing for different parts, which can execute a graph operator efficiently based on the structural patterns in those parts [159]. Such a processing which adapts based on the subgraph structure would be very valuable. Such an adaptive processing can then be separately optimized—either for an algorithmic pattern, or for a specific backend device. In a distributed setup, such a processing relies on an adaptive graph partitioning, which can divide a graph across network nodes in a non-uniform manner, and still achieve an overall performance improvement, due to optimized processing for different subgraph patterns or for a device. Again, this performance could be in terms of execution time, energy efficiency, or other systemic criteria.

8.4.3   Hardware for Graph Processing

Once traversal patterns of graphs are well-understood, it would be possible to cast the common patterns on a chip. This would significantly improve the execution time. Towards this, various primitives such as gather-apply-scatter, pull-apply-push, map-reduce, or even BFS can be hardwired. Graph algorithms can then be modeled using these primitives, which can then be completed with lightening speed. Advances are also required in efficient cache-management for such hardware intrinsics. Further, since the primitives would deal with a large and unknown amount of data, innovative solutions to bridge the gap between processor and memory would be of prime importance [160, 161].

8.4.4 Handling Dynamic Updates

While there have been works supporting dynamic and streaming graph updates [162, 163], focused research on the topic would uncover common patterns across algorithms, and classification of graph algorithms based on these patterns. For instance, incremental processing is computationally easier than the decremental processing in case of shortest paths. In contrast, decremental processing is easier in case of vertex coloring. It is also unclear how to optimize these dynamic updates for different kinds of graphs, or for different subgraphs wherein they are applied. Choosing a backend device for certain kinds of dynamic updates would be an altogether new area of research.

8.4.5 Performance Modeling

A performance model helps predict performance without running a benchmark. Such models have been developed for various specialized programs as well as for different kinds of hardware. It would be helpful to have such a model for graphs on various types of devices [164, 165]. However, due to heavy dependence on both the graph structure and the backend device characteristics, engineering an accurate performance model is a challenge for graphs in a heterogeneous system. Once again, identifying common patterns across algorithms and categorizing graphs across various classes would help create a usable meta-model built on top of good individual device models. Such a model can then be helpful in shared setups such as clouds.

8.4.6 Learning to Aid Graph Processing

Several machine learning algorithms have been cast as graph algorithms, and systems specialized for such a processing have been developed. However, long-running graph algorithms which churn a lot of data, provide a great opportunity to learn about the graph processing itself! Thus, one can learn about patterns of traversals, updates of attributes, or even which devices are bottlenecks for which operations [52, 166, 167]. Such a learning can be very helpful in specializing the processing for a certain kind of graph, a subgraph, or a certain device characteristic. A heterogeneous system handling variety of graphs and running different kinds of algorithms would be an ideal testbed for scheduling graph operations. We expect such a learning on unstructured data to become mainstream as the data continue to grow.

References

1. F. Rahimian, A.H. Payberah, S. Girdzijauskas, S. Haridi, Distributed vertex-cut partitioning, in *Proceedings of the fourth International Conference on Distributed Applications and Interoperable Systems (DAIS).* Lecture Notes in Computer Science, vol. 8460 (2014), pp. 186–200
2. L. Dagum, R. Menon, OpenMP: an industry-standard API for shared-memory programming. IEEE Comput. Sci. Eng. **5**(1), 46–55. https://doi.org/10.1109/99.660313
3. S. Cook, *CUDA Programming: A Developer's Guide to Parallel Computing with GPUs,* 1st edn. (Morgan Kaufmann Publishers, San Francisco, 2013)
4. Thrust: Parallel Programming Library (2009). https://docs.nvidia.com/cuda/thrust/index.html
5. M.P. Forum, MPI: a message-passing interface standard, Tech. Rep., Knoxville, TN, 1994. http://www.ncstrl.org:8900/ncstrl/servlet/search?formname=detail&id=oai%3Ancstrlh%3Autk_cs%3Ancstrl.utk_cs%2F%2FUT-CS-94-230
6. P. Erdős, A. Rènyi, in *On the Evolution of Random Graphs* (Publication of the Mathematical Institute of the Hungarian Academy of Sciences, 1960), pp. 17–61
7. Wikipedia contributors, Scale-free network—Wikipedia, the free encyclopedia (2019). https://en.wikipedia.org/w/index.php?title=Scale-free_network&oldid=930663181. Accessed 17 Dec 2019
8. A.-L. Barabasi, *Linked: How Everything is Connected to Everything Else and What it Means* (Basic Books, New York, 2003)
9. C. Groer, B.D. Sullivan, S. Poole, A mathematical analysis of the R-MAT random graph generator. Networks **58**(3), 159–170 (2011). https://doi.org/10.1002/net.20417
10. U. Meyer, P. Sanders, Delta-stepping: a parallel single source shortest path algorithm, in *Proceedings of the 6th Annual European Symposium on Algorithms, ESA '98* (Springer, London, 1998), pp. 393–404. http://dl.acm.org/citation.cfm?id=647908.740136
11. J.E. Gonzalez, Y. Low, H. Gu, D. Bickson, C. Guestrin, PowerGraph: distributed graph-parallel computation on natural graphs, in *Proceedings of the 10th USENIX Symposium on Operating Systems Design and Implementation, OSDI '12* (2012), pp. 17–30
12. Y. Wang, A. Davidson, Y. Pan, Y. Wu, A. Riffel, J.D. Owens, Gunrock: a high-performance graph processing library on the GPU, in *Proceedings of the 21st ACM SIGPLAN Symposium on Principles and Practice of Parallel Programming (PPoPP) (article 11)* (2016), pp. 1–12
13. L.G. Valiant, A bridging model for parallel computation. Commun. ACM **33**(8), 103–111 (1990)
14. G. Malewicz, M.H. Austern, A.J. Bik, J.C. Dehnert, I. Horn, N. Leiser, G. Czajkowski, Pregel: a system for large-scale graph processing, in *Proceedings of the ACM SIGMOD International Conference on Management of Data* (2010), pp. 135–146

15. A. Ching, S. Edunov, M. Kabiljo, D. Logothetis, S. Muthukrishnan, One trillion edges: graph processing at facebook-scale, in *Proceedings of the VLDB Endowment* (2015), pp. 1804–1815

16. Y. Low, D. Bickson, G.J. Gonzalez, C. Guestrin, A. Kyrola, J.M. Hellerstein, Graphlab: a new parallel framework for machine learning, in *Conference on Uncertainty in Artificial Intelligence (UAI), UAI'10* (AUAI Press, Arlington, 2010), pp. 340–349. http://dl.acm.org/citation.cfm?id=3023549.3023589

17. Y. Low, D. Bickson, J. Gonzalez, C. Guestrin, A. Kyrola, J.M. Hellerstein, Distributed GraphLab: a framework for machine learning and data mining in the cloud, in *Proceedings of the VLDB Endowment* (2012), pp. 716–727

18. M. Burtscher, R. Nasre, K. Pingali, A quantitative study of irregular programs on GPUs, in *IEEE International Symposium on Workload Characterization (IISWC)* (IEEE, Piscataway, 2012), pp. 141–151

19. S. Pai, K. Pingali, A compiler for throughput optimization of graph algorithms on GPUs, in *Proceedings of the 2016 ACM SIGPLAN International Conference on Object-Oriented Programming, Systems, Languages, and Applications, OOPSLA* (ACM, New York, 2016), pp. 1–19

20. A. Gharaibeh, L. Beltrão Costa, E. Santos-Neto, M. Ripeanu, A yoke of oxen and a thousand chickens for heavy lifting graph processing, in *Proceedings of the 21st International Conference on Parallel Architectures and Compilation Techniques, PACT '12* (ACM, New York, 2012), pp. 345–354

21. R. Dathathri, G. Gill, L. Hoang, H.-V. Dang, A. Brooks, N. Dryden, M. Snir, K. Pingali, Gluon: a communication-optimizing substrate for distributed heterogeneous graph analytics, in *Proceedings of the 39th ACM SIGPLAN Conference on Programming Language Design and Implementation (PLDI)* (ACM, New York, 2018), pp. 752–768

22. S. Hong, H. Chafi, E. Sedlar, K. Olukotun, Green-Marl: a DSL for easy and efficient graph analysis, in *Proceedings of the Seventeenth International Conference on Architectural Support for Programming Languages and Operating Systems, ASPLOS XVII* (ACM, New York, 2012), pp. 349–362

23. D. Prountzos, R. Manevich, K. Pingali, Elixir: a system for synthesizing concurrent graph programs, in *Proceedings of the ACM International Conference on Object Oriented Programming Systems Languages and Applications, OOPSLA '12* (ACM, New York, 2012), pp. 375–394. https://doi.org/10.1145/2384616.2384644

24. S. Hong, S. Salihoglu, J. Widom, K. Olukotun, Simplifying scalable graph processing with a domain-specific language, in *Proceedings of Annual IEEE/ACM International Symposium on Code Generation and Optimization, CGO '14* (ACM, New York, 2014), pp. 208–218

25. G. Shashidhar, R. Nasre, LightHouse: an automatic code generator for graph algorithms on GPUs, in *Languages and Compilers for Parallel Computing*, ed. by C. Ding, J. Criswell, P. Wu (Springer, Cham 2017), pp. 235–249

26. U. Cheramangalath, R. Nasre, Y.N. Srikant, Falcon: a graph manipulation language for heterogeneous systems. ACM Trans. Archit. Code Optim. **12**(4), 54:1–54:27 (2015). https://doi.org/10.1145/2842618

27. U. Cheramangalath, R. Nasre, Y.N. Srikant, DH-Falcon: a language for large-scale graph processing on distributed heterogeneous systems, in *2017 IEEE International Conference on Cluster Computing (CLUSTER)* (IEEE, Piscataway, 2017), pp. 439–450

28. E. Moore, in *The Shortest Path Through a Maze*. Bell Telephone System. Technical publications. Monograph (Bell Telephone System, 1959). https://books.google.co.in/books?id=IVZBHAAACAAJ

29. R. Tarjan, Depth-first search and linear graph algorithms. SIAM J. Comput. **1**(2), 146–160 (1972)

30. R.E. Bellman, On a routing problem. Q. Appl. Math. **16**, 87–90 (1958)

31. R.W. Floyd, Algorithm 97: shortest path. Commun. ACM **5**(6), 345–345 (1962). https://doi.org/10.1145/367766.368168

32. T H. Cormen, C. Stein, R.L. Rivest, C.E. Leiserson, *Introduction to Algorithmms*, 2nd edn. (McGraw-Hill, New York 2001)

33. R. Tarjan, Depth-first search and linear graph algorithms. SIAM J. Comput. **1**(2), 146–160 (1972). https://doi.org/10.1137/0201010

34. D. Coppersmith, L. Fleischer, B. Hendrickson, A. Pinar, A divide-and-conquer algorithm for identifying strongly connected components, Tech. Rep., Ernest Orlando Lawrence Berkeley National Laboratory, Berkeley, 2003. https://escholarship.org/uc/item/1hx5n2df

35. R. Prim, Shortest connection networks and some generalizations. Bell System Technol. J. **36**, 1389–1401 (1957)

36. J. Kruskal, On the shortest spanning tree of a graph and the traveling salesman problem. Proc. Am. Math. Soc. **7**, 48–50 (1956)

37. O. Boruvka, O jistem problemu minimalnim(about a certain minimal problem), in *(in czech, germansummary), Prace mor. pnrodoved.*, v(3), 3758 (1926)

38. Wikipedia Contributors, Borůvka's algorithm—wikipedia, the free encyclopedia (2019). https://en.wikipedia.org/w/index.php?title=Bor%C5%AFvka%27s_algorithm&oldid=903222379. Accessed 27 Oct 2019

39. J.J. Whang, A. Lenharth, I.S. Dhillon, K. Pingali, Scalable data-driven pagerank: algorithms, system issues, and lessons learned, in *European Conference on Parallel Processing* (Springer, Berlin, 2015), pp. 438–450

40. Wikipedia Contributors, Pagerank—Wikipedia, the free encyclopedia (2019). https://en.wikipedia.org/w/index.php?title=PageRank&oldid=907975070. Accessed 11 Aug 2019

41. Wikipedia Contributors, Graph coloring—Wikipedia, the free encyclopedia (2019). https://en.wikipedia.org/w/index.php?title=Graph_coloring&oldid=908422192. Accessed 11 Aug 2019

42. Wikipedia Contributors, Greedy coloring—Wikipedia, the free encyclopedia (2019). https://en.wikipedia.org/w/index.php?title=Greedy_coloring&oldid=908906078. Accessed 11 Aug 2019

43. Wikipedia Contributors, Degeneracy (graph theory)—Wikipedia, the free encyclopedia (2019). https://en.wikipedia.org/w/index.php?title=Degeneracy_(graph_theory)&oldid=908159787. Accessed 11 Aug 2019

44. Wikipedia Contributors, Betweenness centrality—Wikipedia, the free encyclopedia (2019). https://en.wikipedia.org/w/index.php?title=Betweenness_centrality&oldid=905972092. Accessed 11 Aug 2019

45. U. Brandes, A faster algorithm for betweenness centrality. J. Math. Soc. **25**(2), 163–177 (2001)

46. D. Chakrabarti, C. Faloutsos, Graph mining: Laws, generators, and algorithms. ACM Comput. Surv. **38**(1) (2006). https://doi.org/10.1145/1132952.1132954

47. S. Parthasarathy, S. Tatikonda, D. Ucar, A survey of graph mining techniques for biological datasets, in *Managing and Mining Graph Data. Advances in Database Systems*, ed. by C. Aggarwal, H. Wang (Springer, Boston, 2010)

48. C. Vicknair, M. Macias, Z. Zhao, X. Nan, Y. Chen, D. Wilkins, A comparison of a graph database and a relational database: a data provenance perspective, in *Proceedings of the 48th Annual Southeast Regional Conference, ACM SE '10* (ACM, New York, 2010), pp. 42:1–42:6. https://doi.org/10.1145/1900008.1900067

49. M. Needham, A.E. Hodler, *Graph Algorithms: Practical Examples in Apache Spark and Neo4j* (O'Reilly Media, Sebastopol, 2019)

50. F. Rousseau, E. Kiagias, M. Vazirgiannis, Text categorization as a graph classification problem, in *Proceedings of the 53rd Annual Meeting of the Association for Computational Linguistics* (2015), pp. 1702–1712

51. F. Rousseau, Graph-of-words: mining and retrieving text with networks of features, Ph.D Thesis, Ecole Polytechnique Laboratoire d'Informatique de l'X (LIX), 2015

52. R. Mihalcea, Graph-based ranking algorithms for sentence extraction, applied to text summarization, in *Proceedings of the ACL 2004 on Interactive Poster and Demonstration Sessions, ACLdemo '04* (Association for Computational Linguistics, Stroudsburg, 2004). http://dx.doi.org/10.3115/1219044.1219064

53. R. Nasre, M. Burtscher, K. Pingali, Atomic-free irregular computations on GPUs, in *Proceedings of the 6th Workshop on General Purpose Processor Using Graphics Processing Units, GPGPU-6* (ACM, New York, 2013), pp. 96–107. https://doi.org/10.1145/2458523.2458533

54. M. Besta, M. Podstawski, L. Groner, E. Solomonik, T. Hoefler, To push or to pull: on reducing communication and synchronization in graph computations, in *Proceedings of the 26th International Symposium on High-Performance Parallel and Distributed Computing, HPDC '17* (ACM, New York, 2017), pp. 93–104. https://doi.org/10.1145/3078597.3078616

55. Y. Xia, V.K. Prasanna, Topologically adaptive parallel breadth-first search on multicore processors, in *Proceedings of 21st Int Conference on Parallel and Distributed Computing Systems, PDCS '09* (2009)

56. A. Aggarwal, R.J. Anderson, M.-Y. Kao, Parallel depth-first search in general directed graphs, in *Proceedings of the Twenty-First Annual ACM Symposium on Theory of Computing, STOC '89* (ACM, New York, 1989), pp. 297–308. https://doi.org/10.1145/73007.73035

57. D. Merrill, M. Garland, A. Grimshaw, Scalable GPU Graph Traversal, in *Proceedings of the 17th ACM SIGPLAN Symposium on Principles and Practice of Parallel Programming, PPoPP '12* (ACM, New York, 2012), pp. 117–128. https://doi.org/10.1145/2145816.2145832

58. A. Kyrola, G. Blelloch, C. Guestrin, GraphChi: large-scale graph computation on just a PC, in *Proceedings of the 10th USENIX Conference on Operating Systems Design and Implementation, OSDI'12* (USENIX Association, Berkeley, 2012), pp. 31–46. http://dl.acm.org/citation.cfm?id=2387880.2387884

59. W. McLendon III, B. Hendrickson, S. Plimpton, L. Rauchwerger, Finding strongly connected components in distributed graphs. J. Parallel Distrib. Comput. **65**(8), 901–910 (2005)

60. S. Hong, N.C. Rodia, K. Olukotun, On fast parallel detection of strongly connected components (SCC) in small-world graphs, in *Proceedings of the International Conference on High Performance Computing, Networking, Storage and Analysis, SC '13* (ACM, New York, 2013), pp. 92:1–92:11. https://doi.org/10.1145/2503210.2503246

61. J. Reif, Depth-first search is inherently sequential. Inf. Process. Lett. **20**(5), 229–234 (1985)

62. J. Shun, L. Dhulipala, G.E. Blelloch, A simple and practical linear-work parallel algorithm for connectivity, in *Proceedings of the International Symposium on Parallel Algorithms and Applications (SPAA)* (2014), pp. 143–153

63. K. Madduri, D. Ediger, K. Jiang, D.A. Bader, D.G. Chavarría-Miranda, A faster parallel algorithm and efficient multithreaded implementations for evaluating betweenness centrality on massive datasets, in *Proceedings of the 23rd IEEE International Symposium on Parallel and Distributed Processing, IPDPS 2009*, Rome, 2009, pp. 1–8

64. R.E. Tarjan, Efficiency of a good but not linear set union algorithm. J. ACM **22**(2), 215–225 (1975). https://doi.org/10.1145/321879.321884

65. R.E. Tarjan, J. van Leeuwen, Worst-case analysis of set union algorithms. J. ACM **31**(2), 245–281 (1984). https://doi.org/10.1145/62.2160

66. K. Pingali, D. Nguyen, M. Kulkarni, M. Burtscher, M.A. Hassaan, R. Kaleem, T.-H. Lee, A. Lenharth, R. Manevich, M. Méndez-Lojo, D. Prountzos, X. Sui, The Tao of parallelism in algorithms. SIGPLAN Not. **46**(6), 12–25 (2011). https://doi.org/10.1145/1993316.1993501

67. R.J. Anderson, H. Woll, Wait-free parallel algorithms for the union-find problem, in *Proceedings of the Twenty-third Annual ACM Symposium on Theory of Computing, STOC '91* (ACM, New York, 1991), pp. 370–380. https://doi.org/10.1145/103418.103458

68. S.V. Jayanti, R.E. Tarjan, A randomized concurrent algorithm for disjoint set union, in *Proceedings of the 2016 ACM Symposium on Principles of Distributed Computing, PODC '16* (ACM, New York, 2016), pp. 75–82. https://doi.org/10.1145/2933057.2933108

69. R. Chen, J. Shi, Y. Chen, B. Zang, H. Guan, H. Chen, Powerlyra: differentiated graph computation and partitioning on skewed graphs. ACM Trans. Parallel Comput. **5**(3), 13:1–13:39 (2019). https://doi.org/10.1145/3298989

70. M. LeBeane, S. Song, R. Panda, J.H. Ryoo, L.K. John, Data partitioning strategies for graph workloads on heterogeneous clusters, in *Proceedings of the International Conference for High Performance Computing, Networking, Storage and Analysis, SC '15* (ACM, New York, 2015), pp. 56:1–56:12. https://doi.org/10.1145/2807591.2807632

71. G. Malewicz, M.H. Austern, A.J. Bik, J.C. Dehnert, I. Horn, N. Leiser, G. Czajkowski, Pregel: a system for large-scale graph processing, in *Proceedings of the 2010 ACM SIGMOD International Conference on Management of Data, SIGMOD '10* (ACM, New York, 2010), pp. 135–146. https://doi.org/10.1145/1807167.1807184

72. Z. Khayyat, K. Awara, A. Alonazi, H. Jamjoom, D. Williams, P. Kalnis, Mizan: a System for dynamic load balancing in large-scale graph processing, in *Proceedings of the 8th ACM European Conference on Computer Systems, EuroSys '13* (ACM, New York, 2013), pp. 169–182. https://doi.org/10.1145/2465351.2465369

73. Y. Jie, T. Guangming, M. Zeyao, S. Ninghui, Graphine: programming graph-parallel computation of large natural graphs for multicore clusters. IEEE Trans. Parallel Distrib. Syst. **27**(6), 1647–1659 (2016)

74. S. Hong, H. Chafi, E. Sedlar, K. Olukotun, Green-Marl: a DSL for easy and efficient graph analysis, in *Proceedings of the Seventeenth International Conference on Architectural Support for Programming Languages and Operating Systems, ASPLOS XVII* (ACM, New York, 2012), pp. 349–362. https://doi.org/10.1145/2150976.2151013

75. D. Prountzos, R. Manevich, K. Pingali, Elixir: a System for synthesizing concurrent graph programs, in *Proceedings of the ACM International Conference on Object Oriented Programming Systems Languages and Applications, OOPSLA '12* (ACM, New York, 2012), pp. 375–394. https://doi.org/10.1145/2384616.2384644

76. Y. Zhang, M. Yang, R. Baghdadi, S. Kamil, J. Shun, S. Amarasinghe, GraphIt: a high-performance graph DSL. Proc. ACM Program. Lang. **2**, 121:1–121:30 (2018). https://doi.org/10.1145/3276491

77. X. Zhu, W. Chen, W. Zheng, X. Ma, Gemini: a computation-centric distributed graph processing system, in *Proceedings of the 12th USENIX Conference on Operating Systems Design and Implementation, OSDI'16* (USENIX Association, Berkeley, 2016), pp. 301–316. http://dl.acm.org/citation.cfm?id=3026877.3026901

78. B. Shao, H. Wang, Y. Li, Trinity: a distributed graph engine on a memory cloud, in *Proceedings of the 2013 ACM SIGMOD International Conference on Management of Data, SIGMOD '13* (ACM, New York, 2013), pp. 505–516. https://doi.org/10.1145/2463676.2467799

79. R. Chen, Y. Yao, P. Wang, K. Zhang, Z. Wang, H. Guan, B. Zang, H. Chen, Replication-based fault-tolerance for large-scale graph processing. IEEE Trans. Parallel Distrib. Syst. **29**(7), 1621–1635 (2018).

80. R. Dathathri, G. Gill, L. Hoang, K. Pingali, Phoenix: a substrate for resilient distributed graph analytics, in *Proceedings of the Twenty-Fourth International Conference on Architectural Support for Programming Languages and Operating Systems, ASPLOS '19* (ACM, New York, 2019), pp. 615–630. https://doi.org/10.1145/3297858.3304056

81. R. Dathathri, G. Gill, L. Hoang, H.-V. Dang, A. Brooks, N. Dryden, M. Snir, K. Pingali, Gluon: a communication-optimizing substrate for distributed heterogeneous graph analytics. SIGPLAN Not. **53**(4), 752–768 (2018). https://doi.org/10.1145/3296979.3192404

82. A. Gharaibeh, L. Beltrão Costa, E. Santos-Neto, M. Ripeanu, A yoke of oxen and a thousand chickens for heavy lifting graph processing, in *Proceedings of the 21st International Conference on Parallel Architectures and Compilation Techniques, PACT '12* (ACM, New York, 2012), pp. 345–354. https://doi.org/10.1145/2370816.2370866

83. Y. Bu, B. Howe, M. Balazinska, M.D. Ernst, HaLoop: efficient iterative data processing on large clusters. Proc. VLDB Endow. **3**(1–2), 285–296 (2010). http://dx.doi.org/10.14778/1920841.1920881

84. P. Harish, P.J. Narayanan, Accelerating large graph algorithms on the GPU Using CUDA, in *Proceedings of the 14th International Conference on High Performance Computing, HiPC'07* (Springer, Berlin, 2007), pp. 197–208. http://dl.acm.org/citation.cfm?id=1782174.1782200

85. P. Harish, V. Vineet, P.J. Narayanan, Large graph algorithms for massively multithreaded architectures, Tech. Rep., 2009

86. M. Burtscher, K. Pingali, An efficient CUDA implementation of the tree-based Barnes hut N-body algorithm, in *GPU Gems*, ed. by W.-M.W. Hwu (Elsevier, Amsterdam, 2011)

87. A.E. Sariyüce, K. Kaya, E. Saule, U.V. Çatalyürek, Betweenness centrality on GPUs and heterogeneous architectures, in *Proceedings of the 6th Workshop on General Purpose Processor Using Graphics Processing Units, GPGPU-6* (ACM, New York, 2013), pp. 76–85. https://doi.org/10.1145/2458523.2458531

88. M. Méndez-Lojo, A. Mathew, K. Pingali, Parallel inclusion-based points-to analysis, in *Proceedings of the ACM International Conference on Object Oriented Programming Systems Languages and Applications, OOPSLA '10* (ACM, New York, 2010), pp. 428–443. https://doi.org/10.1145/1869459.1869495

89. T. Prabhu, S. Ramalingam, M. Might, M. Hall, EigenCFA: accelerating flow analysis with GPUs, in *Proceedings of the 38th Annual ACM SIGPLAN-SIGACT Symposium on Principles of Programming Languages, POPL '11* (ACM, New York, 2011), pp. 511–522. https://doi.org/10.1145/1926385.1926445

90. M. Kulkarni, M. Burtscher, C. Cascaval, K. Pingali, Lonestar: a suite of parallel irregular programs, in *Proceedings of the IEEE International Symposium on Performance Analysis of Systems and Software* (IEEE, Piscataway, 2009), pp. 65–76

91. Z. Jia, Y. Kwon, G. Shipman, P. McCormick, M. Erez, A. Aiken, A distributed multi-GPU system for fast graph processing. Proc. VLDB Endow. **11**(3), 297–310 (2017). https://doi.org/10.14778/3157794.3157799

92. Harshvardhan, A. Fidel, N.M. Amato, L. Rauchwerger, The STAPL parallel graph library, in *Languages and Compilers for Parallel Computing*, ed. by H. Kasahara, K. Kimura (Springer, Berlin, 2013), pp. 46–60

93. Z. Shang, J.X. Yu, Z. Zhang, TuFast: a lightweight parallelization library for graph analytics, in *Proceedings of the 2019 IEEE 35th International Conference on Data Engineering (ICDE)* (IEEE, Piscataway, 2019), pp. 710–721

94. The Boost Graph Library, *User Guide and Reference Manual* (Addison-Wesley Longman Publishing, Boston, 2002)

95. S. Salihoglu, J. Widom, GPS: A graph processing system, in *Proceedings of the 25th International Conference on Scientific and Statistical Database Management, SSDBM* (ACM, New York, 2013), pp. 22:1–22:12. https://doi.org/10.1145/2484838.2484843

96. Y. Bu, V. Borkar, J. Jia, M.J. Carey, T. Condie, Pregelix: Big(Ger) graph analytics on a dataflow engine. Proc. VLDB Endow. **8**(2), 161–172 (2014). http://dx.doi.org/10.14778/2735471.2735477

97. J. Shun, G.E. Blelloch, Ligra: a lightweight graph processing framework for shared memory, in *Proceedings of the 18th ACM SIGPLAN Symposium on Principles and Practice of Parallel Programming, PPoPP '13* (ACM, New York, 2013), pp. 135–146. https://doi.org/10.1145/2442516.2442530

98. J. Shun, L. Dhulipala, G.E. Blelloch, Smaller and faster: parallel processing of compressed graphs with Ligra+, in *2015 Data Compression Conference* (2015), pp. 403–412

99. M. Han, K. Daudjee, Giraph unchained: barrierless asynchronous parallel execution in Pregel-like graph processing systems. Proc. VLDB Endow. **8**(9), 950–961 (2015). https://doi.org/10.14778/2777598.2777604

100. K. Siddique, Z. Akhtar, Y. Kim, Y.-S. Jeong, E.J. Yoon, Investigating Apache Hama: a bulk synchronous parallel computing framework. J. Supercomput. **73**(9), 4190–4205 (2017). https://doi.org/10.1007/s11227-017-1987-9

101. K. Lee, L. Liu, K. Schwan, C. Pu, Q. Zhang, Y. Zhou, E. Yigitoglu, P. Yuan, Scaling iterative graph computations with GraphMap, in *Proceedings of the International Conference for High Performance Computing, Networking, Storage and Analysis, SC '15* (2015), pp. 1–12

102. Y. Cheng, F. Wang, H. Jiang, Y. Hua, D. Feng, Z. Wang, LCC-graph: a high-performance graph-processing framework with low communication costs, in *2016 IEEE/ACM 24th International Symposium on Quality of Service (IWQoS)* (2016), pp. 1–10

103. T. White, *Hadoop: The Definitive Guide*, 1st edn. (O'Reilly Media, Sebastopol, 2009)

104. J. Ekanayake, H. Li, B. Zhang, T. Gunarathne, S.-H. Bae, J. Qiu, G. Fox, Twister: a runtime for iterative mapreduce, in *Proceedings of the 19th ACM International Symposium on High Performance Distributed Computing, HPDC '10* (ACM, New York, 2010), pp. 810–818. https://doi.org/10.1145/1851476.1851593

105. P.A. Bernstein, V. Hadzilacos, N. Goodman, *Concurrency Control and Recovery in Database Systems* (Addison-Wesley, Reading, 1987). http://research.microsoft.com/en-us/people/philbe/ccontrol.aspx

106. M. Herlihy, J.E.B. Moss, Transactional memory: architectural support for lock-free data structures, in *Proceedings of the 20th Annual International Symposium on Computer Architecture (ISCA)*, San Diego, 1993, pp. 289–300

107. K. Vora, S.C. Koduru, R. Gupta, ASPIRE: exploiting asynchronous parallelism in iterative algorithms using a relaxed consistency based DSM, in *Proceedings of the 2014 ACM International Conference on Object Oriented Programming Systems Languages and Applications, OOPSLA '14* (ACM, New York, 2014), pp. 861–878. https://doi.org/10.1145/2660193.2660227

108. Harshvardhan, A. Fidel, N.M. Amato, L. Rauchwerger, KLA: a new algorithmic paradigm for parallel graph computations, in *Proceedings of the 23rd International Conference on Parallel Architectures and Compilation, PACT '14* (ACM, New York, 2014), pp. 27–38. https://doi.org/10.1145/2628071.2628091

109. K. Vora, C. Tian, R. Gupta, Z. Hu, CoRAL: confined recovery in distributed asynchronous graph processing, in *Proceedings of the Twenty-Second International Conference on Architectural Support for Programming Languages and Operating Systems, ASPLOS '17* (ACM, New York, 2017), pp. 223–236. https://doi.org/10.1145/3037697.3037747

110. G. Wang, W. Xie, A. Demers, J. Gehrke, Asynchronous large-scale graph processing made easy, in *Proceedings of the Sixth Biennial Conference on Innovative Data Systems Research (CIDR'13)* (2013)

111. A. Kyrola, J. Shun, G. Blelloch, Beyond synchronous: new techniques for external-memory graph connectivity and minimum spanning forest, in *Experimental Algorithms*, ed. by J. Gudmundsson, J. Katajainen (Springer, Cham, 2014), pp. 123–137

112. W.-S. Han, S. Lee, K. Park, J.-H. Lee, M.-S. Kim, J. Kim, H. Yu, Turbograph: a fast parallel graph engine handling billion-scale graphs in a single PC, in *Proceedings of the 19th ACM SIGKDD International Conference on Knowledge Discovery and Data Mining, KDD '13* (ACM, New York, 2013), pp. 77–85. https://doi.org/10.1145/2487575.2487581

113. S. Ko, W.-S. Han, Turbograph++: a scalable and fast graph analytics system, in *Proceedings of the 2018 International Conference on Management of Data, SIGMOD '18* (ACM, New York, 2018), pp. 395–410. https://doi.org/10.1145/3183713.3196915

114. J.E. Gonzalez, Y. Low, H. Gu, D. Bickson, C. Guestrin, PowerGraph: distributed graph-parallel computation on natural graphs, in *Proceedings of the 10th USENIX Conference on Operating Systems Design and Implementation, OSDI'12* (USENIX Association, Berkeley, 2012), pp. 17–30. http://dl.acm.org/citation.cfm?id=2387880.2387883

115. J. Wu, R. Das, J. Saltz, H. Berryman, S. Hiranandan, Distributed memory compiler design for sparse problems. IEEE Trans. Comput **44**(6), 737–753 (1995)

116. Y. Wang, A. Davidson, Y. Pan, Y. Wu, A. Riffel, J.D. Owens, Gunrock: a high-performance graph processing library on the GPU, in *Proceedings of the 21st ACM SIGPLAN Symposium on Principles and Practice of Parallel Programming, PPoPP '16* (ACM, New York, 2016), pp. 11:1–11:12. https://doi.org/10.1145/2851141.2851145

117. S. Xiao, W.C. Feng, Inter-block GPU communication via fast barrier synchronization, in *2010 IEEE International Symposium on Parallel Distributed Processing (IPDPS)* (2010), pp. 1–12

118. Q. Xu, H. Jeon, M. Annavaram, Graph processing on GPUs: where are the bottlenecks? in *2014 IEEE International Symposium on Workload Characterization (IISWC)* (2014), pp. 140–149

119. Y. Perez, R. Sosič, A. Banerjee, R. Puttagunta, M. Raison, P. Shah, J. Leskovec, Ringo: interactive graph analytics on big-memory machines, in *Proceedings of the 2015 ACM SIGMOD International Conference on Management of Data, SIGMOD '15* (ACM, New York, 2015), pp. 1105–1110. https://doi.org/10.1145/2723372.2735369

120. S. Pai, K. Pingali, A compiler for throughput optimization of graph algorithms on GPUs, in *Proceedings of the 2016 ACM SIGPLAN International Conference on Object-Oriented Programming, Systems, Languages, and Applications, OOPSLA 2016* (ACM, New York, 2016), pp. 1–19. https://doi.org/10.1145/2983990.2984015

121. A. Davidson, S. Baxter, M. Garland, J.D. Owens, Work-efficient parallel GPU methods for single source shortest paths, in *Proceedings of the 2014 IEEE 28th International Symposium on Parallel and Distributed Processing IPDPS 2014* (IEEE, Piscataway, 2014)

122. R. Nasre, M. Burtscher, K. Pingali, Morph algorithms on GPUs, in *Proceedings of the 18th ACM SIGPLAN Symposium on Principles and Practice of Parallel Programming, PPoPP '13* (ACM, New York, 2013), pp. 147–156. https://doi.org/10.1145/2442516.2442531

123. Z. Fu, M. Personick, B. Thompson, MapGraph: a high level API for fast development of high performance graph analytics on GPUs, in *Proceedings of Workshop on GRAph Data Management Experiences and Systems, GRADES'14* (ACM, New York, 2014), pp. 2:1–2:6. https://doi.org/10.1145/2621934.2621936

124. P. Zhao, X. Luo, J. Xiao, X. Shi, H. Jin, Puffin: graph processing system on multi-GPUs, in *2017 IEEE 10th Conference on Service-Oriented Computing and Applications (SOCA)* (IEEE, Piscataway, 2017), pp. 50–57

125. A.H. Nodehi Sabet, J. Qiu, Z. Zhao, Tigr: transforming irregular graphs for GPU-friendly graph processing, in *Proceedings of the Twenty-Third International Conference on Architectural Support for Programming Languages and Operating Systems, ASPLOS '18* (ACM, New York, 2018), pp. 622–636. https://doi.org/10.1145/3173162.3173180

126. P. Zhang, M. Zalewski, A. Lumsdaine, S. Misurda, S. McMillan, GBTL-CUDA: graph algorithms and primitives for GPUs, in *2016 IEEE International Parallel and Distributed Processing Symposium Workshops (IPDPSW)* (2016), pp. 912–920

127. D. Sengupta, S.L. Song, K. Agarwal, K. Schwan, GraphReduce: processing large-scale graphs on accelerator-based systems, in *Proceedings of the International Conference for High Performance Computing, Networking, Storage and Analysis, SC '15* (2015), pp. 1–12

128. W. Zhong, J. Sun, H. Chen, J. Xiao, Z. Chen, C. Cheng, X. Shi, Optimizing graph processing on GPUs. IEEE Trans Parallel Distrib. Syst. **28**(4), 1149–1162 (2017)

129. J. Zhong, B. He, Medusa: Simplified Graph Processing on GPUs. IEEE Trans. Parallel Distrib. Syst. **25**(6), 1543–1552 (2014). https://doi.org/10.1109/TPDS.2013.111

130. A. Aiken, M. Bauer, S. Treichler, Realm: an event-based low-level runtime for distributed memory architectures, in *2014 23rd International Conference on Parallel Architecture and Compilation Techniques (PACT)* (2014), pp. 263–275

131. G. Gill, R. Dathathri, L. Hoang, K. Pingali, A study of partitioning policies for graph analytics on large-scale distributed platforms. Proc. VLDB Endow. **12**(4), 321–334 (2018). https://doi.org/10.14778/3297753.3297754

132. Y. Simmhan, A. Kumbhare, C. Wickramaarachchi, S. Nagarkar, S. Ravi, C. Raghavendra, V. Prasanna, Goffish: a sub-graph centric framework for large-scale graph analytics, in *Euro-Par 2014 Parallel Processing*, ed. by F. Silva, I. Dutra, V. Santos Costa (Springer, Cham, 2014), pp. 451–462

133. J.E. Gonzalez, R.S. Xin, A. Dave, D. Crankshaw, M. J. Franklin, I. Stoica, GraphX: graph processing in a distributed dataflow framework, in *Proceedings of the 11th USENIX Conference on Operating Systems Design and Implementation, OSDI'14* (USENIX Association, Berkeley, 2014), pp. 599–613. http://dl.acm.org/citation.cfm?id=2685048.2685096

134. M. Pundir, L. M. Leslie, I. Gupta, R.H. Campbell, Zorro: zero-cost reactive failure recovery in distributed graph processing, in *Proceedings of the Sixth ACM Symposium on Cloud Computing, SoCC '15* (ACM, New York, 2015), pp. 195–208. https://doi.org/10.1145/2806777.2806934

135. S. Hong, S. Depner, T. Manhardt, J. Van Der Lugt, M. Verstraaten, H. Chafi, PGX.D: a fast distributed graph processing engine, in *Proceedings of the International Conference for High Performance Computing, Networking, Storage and Analysis, SC '15* (IEEE, Piscataway, 2015), pp. 1–12

136. D. Yan, J. Cheng, Y. Lu, W. Ng, Blogel: a block-centric framework for distributed computation on real-world graphs. Proc. VLDB Endow. **7**(14), 1981–1992 (2014). https://doi.org/10.14778/2733085.2733103

137. S.M. Faisal, S. Parthasarathy, P. Sadayappan, Global graphs: a middleware for large scale graph processing, in *2014 IEEE International Conference on Big Data (Big Data)* (IEEE, Piscataway, 2014), pp. 33–40

138. Y. Zhao, K. Yoshigoe, M. Xie, S. Zhou, R. Seker, J. Bian, LightGraph: lighten communication in distributed graph-parallel processing, in *2014 IEEE International Congress on Big Data* (IEEE, Piscataway, 2014), pp. 717–724

139. R. Dathathri, G. Gill, L. Hoang, H.-V. Dang, A. Brooks, N. Dryden, M. Snir, K. Pingali, Gluon: a communication-optimizing substrate for distributed heterogeneous graph analytics, in *Proceedings of the 39th ACM SIGPLAN Conference on Programming Language Design and Implementation, PLDI 2018* (ACM, New York, 2018), pp. 752–768. https://doi.org/10.1145/3192366.3192404

140. R. Dathathri, G. Gill, L. Hoang, K. Pingali, Phoenix: a substrate for resilient distributed graph analytics, in *Proceedings of the Twenty-Fourth International Conference on Architectural Support for Programming Languages and Operating Systems* (ACM, New York, 2019), pp. 615–630. https://doi.org/10.1145/3297858.3304056

141. R. Dathathri, G. Gill, L. Hoang, V. Jatala, K. Pingali, V.K. Nandivada, H. Dang, M. Snir, Gluon-Async: a bulk-asynchronous system for distributed and heterogeneous graph analytics, in *2019 28th International Conference on Parallel Architectures and Compilation Techniques (PACT)* (2019), pp. 15–28

142. M. Flynn, Some computer organizations and their effectiveness. IEEE Trans. Comput. **100**(9), 948–960 (1972)

143. Nvidia GK-110B Specifications (2013). https://www.nvidia.com/content/dam/en-zz/Solutions/Data-Center/tesla-product-literature/TeslaK80-datasheet.pdf

144. Nvidia GTX-870M Specifications (2014). https://www.geforce.com/hardware/notebook-gpus/geforce-gtx-870m/specifications

145. Nvidia GTX-1080 Specifications (2017). https://images-eu.ssl-images-amazon.com/images/I/91las2p%2BDnS.pdf

146. Nvidia GK-110B Specifications (2018). https://www.nvidia.com/content/dam/en-zz/Solutions/design-visualization/productspage/quadro/quadro-desktop/quadro-volta-gv100-data-sheet-us-nvidia-704619-r3-web.pdf

147. A. Coates, B. Huval, T. Wang, D.J. Wu, A.Y. Ng, B. Catanzaro, Deep learning with COTS HPC systems, in *Proceedings of the 30th International Conference on International Conference on Machine Learning, ICML'13*, vol. 28 (2013), pp. III-1337–III-1345. JMLR.org, http://dl.acm.org/citation.cfm?id=3042817.3043086

148. A. Eklund, P. Dufort, D. Forsberg, S.M. LaConte, Medical image processing on the GPU—past, present and future. Med. Image Anal. **17**(8), 1073–1094

149. Y. Go, M. Jamshed, Y. Moon, C. Hwang, K. Park, APUNet: revitalizing GPU as packet processing accelerator, in *Proceedings of the 14th USENIX Conference on Networked Systems Design and Implementation, NSDI'17* (USENIX Association, Berkeley, 2017), pp. 83–96. http://dl.acm.org/citation.cfm?id=3154630.3154638

150. M. Thorup, Dynamic graph algorithms with applications, in *Proceedings of the 7th Scandinavian Workshop on Algorithm Theory, SWAT '00* (Springer, Berlin, 2000), pp. 1–9. http://dl.acm.org/citation.cfm?id=645900.672593

151. G. Ramalingam, T. Reps, On the computational complexity of dynamic graph problems. Theor. Comput. Sci. **158**(1–2), 233–277 (1996). https://doi.org/10.1016/0304-3975(95)00079-8

152. L.S. Buriol, M.G.C. Resende, M. Thorup, Speeding up dynamic shortest-path algorithms. INFORMS J. Comput. **20**(2), 191–204 (2008). https://doi.org/10.1287/ijoc.1070.0231

153. S.-W. Cheng, T.K. Dey, J.R. Shewchuk, *Delaunay Mesh Generation* (CRC Press, Boca Raton, 2012)

154. D.A. Bader, K. Madduri, *GTgraph: A Synthetic Graph Generator Suite*, Atlanta, 2006

155. *Ninth DIMACS Implementation Challenge—Shortest Paths* (2006). http://www.dis.uniroma1.it/challenge9/download.shtml

156. P. Boldi, M. Santini, S. Vigna, A large time-aware web graph. SIGIR Forum **42**(2), 33–38 (2008). https://doi.org/10.1145/1480506.1480511

157. H. Kwak, C. Lee, H. Park, S. Moon, What is twitter, a social network or a news media?, in *Proceedings of the 19th International Conference on World Wide Web, WWW '10* (ACM, New York, 2010), pp. 591–600. https://doi.org/10.1145/1772690.1772751

158. B. Gogoi, U. Cheramangalath, R. Nasre, Custom code generation for a graph DSL, in *Proceedings of the 13th Annual Workshop on General Purpose Processing Using Graphics Processing Unit, GPGPU'20* (ACM, New York, 2020), pp. 51–60. https://doi.org/10.1145/3366428.3380772

159. R. Chen, J. Shi, Y. Chen, H. Chen, PowerLyra: differentiated graph computation and partitioning on skewed graphs, in *Proceedings of the Tenth European Conference on Computer Systems, EuroSys '15* (ACM, New York, 2015), pp. 1:1–1:15. https://doi.org/10.1145/2741948.2741970

160. A. Mukkara, N. Beckmann, M. Abeydeera, X. Ma, D. Sanchez, Exploiting locality in graph analytics through hardware-accelerated traversal scheduling, in *2018 51st Annual IEEE/ACM International Symposium on Microarchitecture (MICRO)* (2018), pp. 1–14

161. S. Khoram, J. Zhang, M. Strange, J. Li, Accelerating graph analytics by co-optimizing storage and access on an FPGA-HMC platform, in *Proceedings of the 2018 CM/SIGDA International Symposium on Field-Programmable Gate Arrays, FPGA '18* (Association for Computing Machinery, New York, 2018), pp. 239–248. https://doi.org/10.1145/3174243.3174260

162. G. Malhotra, H. Chappidi, R. Nasre, Fast dynamic graph algorithms, in *Languages and Compilers for Parallel Computing*, ed. by L. Rauchwerger (Springer, Cham, 2017), pp. 262–277

163. U.A. Acar, D. Anderson, G.E. Blelloch, L. Dhulipala, Parallel batch-dynamic graph connectivity, in *The 31st ACM Symposium on Parallelism in Algorithms and Architectures, SPAA '19* (Association for Computing Machinery, New York, 2019), pp. 381–392. https://doi.org/10.1145/3323165.3323196

164. Z. Li, B. Zhang, S. Ren, Y. Liu, Z. Qin, R.S.M. Goh, M. Gurusamy, Performance modelling and cost effective execution for distributed graph processing on configurable VMs, in *Proceedings of the 17th IEEE/ACM International Symposium on Cluster, Cloud and Grid Computing, CCGrid '17* (IEEE Press, New York, 2017), pp. 74–83. https://doi.org/10.1109/CCGRID.2017.85

165. S.D. Pollard, S. Srinivasan, B. Norris, A Performance and recommendation system for parallel graph processing implementations: work-in-progress, in *Companion of the 2019 ACM/SPEC International Conference on Performance Engineering, ICPE '19* (Association for Computing Machinery, New York, 2019), pp. 25–28. https://doi.org/10.1145/3302541.3313097

166. T.D. Bui, S. Ravi, V. Ramavajjala, Neural graph learning: training neural networks using graphs, in *Proceedings of the Eleventh ACM International Conference on Web Search and Data Mining, WSDM '18* (Association for Computing Machinery, New York, 2018), pp. 64–71. https://doi.org/10.1145/3159652.3159731

167. R. Al-Rfou, B. Perozzi, D. Zelle, DDGK: learning graph representations for deep divergence graph kernels, in *The World Wide Web Conference, WWW '19* (Association for Computing Machinery, New York, 2019), pp. 37–48. https://doi.org/10.1145/3308558.3313668

Index

© Springer Nature Switzerland AG 2020
U. Cheramangalath et al., *Distributed Graph Analytics*,
https://doi.org/10.1007/978-3-030-41886-1

Printed in the United States
by Baker & Taylor Publisher Services